CatchUp

Biology 2nd Edition

for the medical sciences

Biology 2nd Edition

for the medical sciences

Philip Bradley and Jane Calvert

Faculty of Medical Sciences, The Medical School, Newcastle University, Framlington Place, Newcastle upon Tyne, UK

Scion

© Scion Publishing Ltd, 2013

Second edition first published 2013

First edition published 2006, reprinted 2009, 2010, 2011, 2012

A CIP catalogue record for this book is available from the British Library.

ISBN 978 1 904842 88 0

Scion Publishing Limited
The Old Hayloft, Vantage Business Park, Bloxham Rd, Banbury, OX16 9UX, UK
www.scionpublishing.com

Important Note from the Publisher

Typeset by Phoenix Photosetting, Chatham, Kent, UK
Printed in the UK by 4edge Ltd, Hockley, Essex.

Contents

Preface

Students entering university courses in the medical or biomedical sciences have a wide range of different qualifications and knowledge. Depending on the route of entry, different students will have covered topics in varying levels of detail. This short text aims to provide an overview of some of the important concepts that will help a student to understand and gain maximum benefit from their university course.

The book takes a hierarchical approach, starting with an introduction to the molecules of life, moving on to consider cells and their functions, how cells are assembled into tissues and ultimately the various systems of the body. Biology is a huge subject so this text selects material that will be most useful to students studying courses related to medicine and the medical sciences. In preparing this Second Edition we have listened to feedback from students and lecturers as to what additional topics would be useful and, as a result, we have included new material on genetic disease (Chapter 13) and have expanded the section on microorganisms to become a separate chapter (Chapter 7). There are a number of other updates in the text which reflect changing concepts regarding cellular organisation.

It is important to be aware of the position of humans in relation to other life on the planet. Life can be divided up into prokaryotes and eukaryotes, which differ markedly in the properties of their cells (Chapter 6). The prokaryotes include two major domains: the bacteria and the archaea, a detailed discussion of which is beyond the scope of this book. The eukaryotes can be divided into kingdoms, as shown in the diagram below.

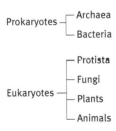

Prokaryotes — Archaea
            — Bacteria

Eukaryotes — Protista
           — Fungi
           — Plants
           — Animals

Classification of humans within the animal kingdom

Phylum	Chordata
Class	Mammalia
Order	Primates
Family	Hominidae
Genus	Homo
Species	*Homo sapiens*

Humans, of course, belong to the animal kingdom, but this contains many thousands of species and so we further classify humans as shown. Biological classification groups together organisms according to degrees of similarity and attempts to reflect evolutionary relationships. The material covered in the first part of the book is relevant to the whole of biology, because the molecules of life are, to a very large extent, shared across the kingdoms. As we go further into the book the material covered becomes more selective. Humans are mammals and, whilst the focus here is on human biology, most of the information covered will be true for other mammalian species.

Biology is a fascinating subject because it tells us how our bodies work and helps us to understand what can go wrong in disease. It is also a subject that has progressed in leaps and bounds as new technologies have allowed us to analyse the processes of life at ever more sophisticated levels. We hope that you will continue to be excited by this science and share your enthusiasm with others.

Philip Bradley and Jane Calvert
March 2013

Acknowledgements

We would like to thank all those who have helped and advised in relation to the material in this book, including both reviewers and our academic colleagues. Most particularly, we would like to thank Austin Diamond and Monica Hughes for their comments on sections of the book. We would also like to thank Julie Alexander for her patient help and support.

Most of all we are grateful to all our many students over the years who have taught us far more than we taught them.

01 Water and life

> **BASIC CONCEPTS:**
> Water makes up approximately 60% of the human body. Its molecular structure allows it to act as a solvent for many of the other key molecules which enable cells to function and life to be maintained. An understanding of the distribution of water in the body, the composition of the various fluid compartments and the control of the movement of water between compartments is crucial to understanding many basic life processes.

1.1 The properties of water

Water is essential for life. The cells of living organisms are composed of around 70% water and many of the reactions essential to life occur in an aqueous environment. The chemical properties of water make it a particularly suitable medium for supporting life. Water is a *polar* molecule, which is to say it has an uneven distribution of charge (Fig. 1.1).

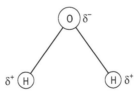

Figure 1.1. A water molecule showing distribution of charge

This means that it is able to interact with other polar and charged groups. Molecules or groups that interact with water are described as *hydrophilic*, whereas non-polar groups are described as *hydrophobic*.

Virtually all the molecules of life are based around the element carbon. These include:

- sugars and polysaccharides
- amino acids and proteins
- nucleotides and nucleic acids
- lipids

Polysaccharides, proteins and nucleic acids are very large molecules, termed *macromolecules*, and are polymers of sugars, amino acids and nucleotides respectively. Biological macromolecules contain both hydrophilic groups (such as OH, NH_2 and COOH) and hydrophobic groups (for example hydrocarbons) and the relative amounts of these influence solubility (for further information see Section 3.7 in *Catch Up Chemistry*).

Interactions with water play an important part in determining the structure of these biological molecules. Generally speaking, hydrophilic groups tend to be exposed on the surface of a molecule or structure from where they are able to interact with water molecules. In contrast, hydrophobic groups tend to orientate themselves towards the inside of the molecule or structure where they interact with each other forming hydrophobic bonds. Interactions between hydrophobic chains of fatty acids allow the formation of cell membranes (see Chapter 4). Other molecules that are associated with membranes, such as proteins, often have hydrophobic regions which are inserted into the membrane to form an anchor.

Water is also very important as a medium of transport and forms the basis of blood. Gases dissolve in water, and this is important in allowing oxygen to be taken to cells and carbon dioxide to be removed.

1.2 Water in the human body

Approximately 60% of the weight of the human body is water – thus a 60 kg person will contain approximately 36 litres of water. Within the body the water is distributed between three main compartments. The bulk of body water (65%) is contained in the cytoplasm of cells and is known as *intracellular fluid*. Most of the remaining extracellular fluid is divided into the *interstitial fluid* (25%) which bathes the cells and the *plasma* (7.5%) which is contained within the blood vessels of the circulatory system. The remaining 2.5% of fluid is known as *transcellular fluid* and includes, for example, the water in the bladder and the contents of the gastrointestinal tract.

Intracellular fluid is separated from interstitial fluid by the plasma membrane of the cell (see Chapter 6). The ionic composition of these two compartments is dramatically different. The extracellular fluid has a similar composition to seawater and contains approximately 140 mmol Na^+ and 110 mmol Cl^-. Extracellular fluid also contains significant levels of bicarbonate ions. By contrast, intracellular fluid contains high levels of K^+ (approximately 160 mmol compared with 4 mmol in extracellular fluid) and low levels of Na^+ (10 mmol). The intracellular negative charge is provided not by Cl^- but by proteins, bicarbonate and phosphate ions.

The concentration gradients of Na^+ and K^+ across cell membranes form the basis of many physiological processes (see Chapters 9 and 22). Ions contained within body fluids are known as *electrolytes*.

A general rule which applies when considering the ionic balance of any one compartment is that it should contain equivalent positive and negative charges (determined by the relative numbers of cations and anions). Each compartment is said to be *electroneutral*. This has significance when considering the movement of ions across membranes because, wherever possible, the body strives to ensure that movement of positively charged cations is accompanied by an equivalent negative charge in anions. When this does not happen electrical potentials are generated across membranes and this forms the basis of the function of excitable tissues (see Chapter 16).

The two components of extracellular fluid are separated from each other by the capillary wall. In most capillaries this is freely permeable to the movement of ions and small organic molecules but does not allow the passage of proteins. Thus under normal circumstances interstitial fluid contains no protein whereas both plasma and intracellular fluid are protein rich.

CLINICAL EXAMPLE: DEHYDRATION

On a hot day a runner may lose up to 2 litres per hour in sweat. For a normal individual a loss of water constituting more than 3% of body weight (about 2 litres) may lead to the early stages of clinical dehydration and cause feelings of light-headedness and disorientation. Further water loss will affect the ability of cells to function and may lead to death due to shock caused by low blood volume. This is why it is particularly important for fun runners to ensure that they take on plenty of water when competing in marathons and other long-distance races.

1.3 Test yourself

The answers are given on p. 175.

Question **1.1**
Where in a biological macromolecule would hydrophobic groups generally be found?

Question **1.2**
What are the three main compartments in which body water is distributed?

Question **1.3**
What is the main cation of: (a) extracellular fluid; (b) intracellular fluid?

Question **1.4**
Organic molecules are based around which element?

Question **1.5**
Which key component of plasma does not normally pass across the capillary wall?

02 Proteins

> **BASIC CONCEPTS:**
>
> Proteins are macromolecules assembled as a sequence of amino acids. There are twenty different amino acids, giving rise to a wide range of possible proteins. According to the particular amino acid sequence, proteins will adopt different three-dimensional structures. Proteins are present in all cells and can perform many roles, including as structural elements and as enzymes. It is important to understand how the amino acid sequence of proteins can determine the properties of different proteins, and also how these properties can be altered by external factors such as the binding of another molecule or the addition of a phosphate group.

2.1 Introduction

Proteins are a highly diverse and important group of molecules, central to life. Proteins are biological macromolecules and are polymers of *amino acids*.

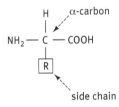

Figure 2.1. General structure of an amino acid

Amino acids contain an amino group and a carboxylic acid group (Fig. 2.1), both attached to an alpha carbon atom. Also attached to the alpha carbon is a side chain, which is different in different amino acids (Fig. 2.2). Side chains have their properties too – some carry a positive or negative charge, some are polar and others are hydrophobic (they prefer not to be in contact with water). The different properties of the side chains are important in determining the structure and function of proteins. There are twenty different amino acids that are found in proteins. Because these can occur in different orders and combinations, this leads to a very large number of possible protein structures.

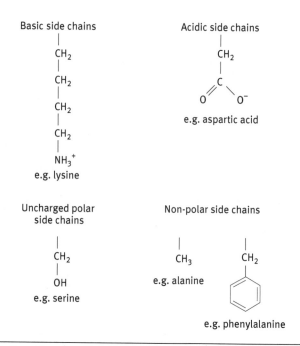

Figure 2.2. Examples of different kinds of amino acids

Amino acids can exist as different isomers, depending upon the arrangement of the groups attached to the alpha carbon. Isomers are defined as 'two or more different compounds with the same chemical formula but different structures and characteristics'. The alpha carbon in an amino acid participates in four covalent bonds forming a tetrahedral arrangement, and mirror image forms can exist, called *enantiomers*. The different enantiomers are described by the letters D and L. All amino acids occurring in proteins are L-isomers.

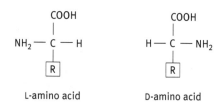

Figure 2.3. D and L forms of amino acids

Amino acids are joined together by *peptide bonds* (Fig. 2.4). A peptide bond is formed in a reaction between the carboxylic acid group of one amino acid and the amino group of another. In the process, a molecule of water is lost and so this is called a *condensation reaction*.

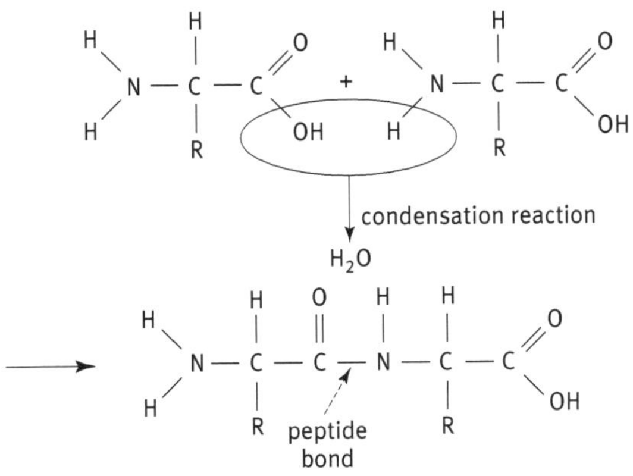

Figure 2.4. Formation of a peptide bond between two amino acids

The amino acids at each end of a protein molecule participate in only one peptide bond, hence they have either a free NH_2 group or a free COOH group. The end of the polypeptide chain with a free amino group is called the *N-terminus*, and the end with the free carboxyl group is called the *C-terminus*.

2.2 Primary structure

Each protein has its own unique amino acid sequence. The sequence of amino acids in a protein defines its primary structure and this sequence is encoded by the gene for the protein.

Depending on the amino acid sequence, proteins will, under physiological conditions, preferentially adopt a particular folded structure, or *conformation* (see the sections on secondary and tertiary structure below). The conformation of the protein is maintained by non-covalent interactions involving amino acid side chains. These include ionic bonds between positive and negatively charged amino acid residues, hydrogen bonds, van der Waals forces and hydrophobic interactions (see *Catch Up Chemistry* for further information on these). Hydrophobic interactions are particularly important as they bring together non-polar amino acid side chains and ensure that these are not exposed to water. Following the initial synthesis of proteins within cells (see Chapter 5) their folding into secondary and tertiary structures is aided by the presence of other proteins called *molecular chaperones*.

2.3 Secondary structures

Some common patterns of folding occur. Two classical protein folds that recur in many different proteins are the *alpha helix* and the *beta sheet*

(Fig. 2.5). Both of these structures depend on interactions between groups in the polypeptide backbone and they can be found in a wide range of protein molecules.

Alpha helix

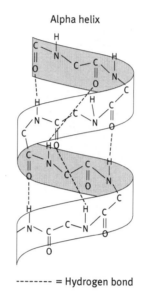

-------- = Hydrogen bond

Beta sheet

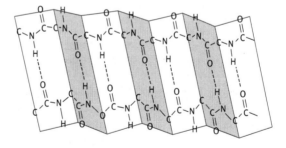

Figure 2.5. An alpha helix and a beta sheet structure

Alpha helix

This is a structure in which the polypeptide chain twists on itself in a highly regular manner. The structure is stabilised by hydrogen bonding between the C=O group of one amino acid residue and the NH group of the residue four amino acids further along the primary sequence. In this way every C=O and

NH, as well as being involved in the covalent peptide bond, also participates in hydrogen bonding. This rule confers particular dimensions on the alpha helix – each turn of the helix represents 3.6 amino acid residues. Alpha helices normally assume a right-handed or clockwise twist as this is energetically more favourable. The amino acid side chains are exposed on the outside of the helix. Some amino acids, in particular proline, tend to disrupt an alpha helical structure and are known as *helix breakers*.

Two or more alpha helices can intertwine to form a *superhelix*. Such superhelices are found in proteins such as keratin, the major constituent of hair.

Beta sheet

The beta sheet, like the alpha helix, is a structure which is maintained by hydrogen bonding between C=O and NH groups, but in this case these interactions occur between adjacent strands. The amino acid side chains protrude alternately above and below the plane of the sheet. Where the adjacent strands run in a similar orientation this is said to be a *parallel* beta sheet; where the adjacent strands lie in opposing orientations the sheet is said to be *anti-parallel*. The beta-sheet structure is found in many important proteins, including antibodies.

Collagen triple helix

A third type of structure is found in collagen, an important structural protein of connective tissues. Collagen has a triple-helical structure, in which three amino acid chains are wound around each other. This structure is allowed because of the particular primary sequence of collagen and related proteins – every third amino acid residue is a glycine. Because glycine (Gly) has the smallest side chain (H) this allows the chains to interact closely to form the triple helix. Collagen-like proteins are also rich in the amino acid proline and show a repeat sequence of Gly–X–Y. Although X and Y can be any amino acid, these positions are most commonly taken by proline and hydroxyproline.

2.4 Tertiary structure

In aqueous solution, all proteins fold in such a way as to internalise those amino acid side chains that are hydrophobic and to ensure that those that are exposed to the solvent are charged or polar residues. Many alpha helices and beta-sheet structures are found to be *amphipathic* – this means that the hydrophobic side chains tend to occur on one side of the helix or beta sheet and the polar and charged residues on the other. In this way the protein can fold to ensure that hydrophobic surfaces interact with each other, and not with solvent.

> **CLINICAL EXAMPLE: BSE AND CJD**
>
> It is crucial to cellular function that proteins fold into their correct conformation. Some diseases are associated with the presence of abnormally folded proteins. The prion disease bovine spongiform encephalopathy (BSE) and its human equivalent Creutzfeldt–Jakob disease (CJD) are caused by the prion protein PrP, which differs from a normal cellular protein by only a few amino acids. The prion protein, in which α-helices are replaced with β-sheets, assumes an abnormal conformation which is resistant to digestion by proteases (the enzymes involved in the breakdown of proteins). Moreover, in the presence of the prion protein, the normal cellular form of the protein is induced to adopt a similar conformation. The presence of these abnormally folded proteins is associated with degeneration of the brain, dementia and death.

2.5 Quaternary structure

A protein can comprise a single polypeptide chain or multiple polypeptides which are associated by covalent and/or non-covalent bonds. A common type of covalent bond is the *disulphide bridge*, which can occur between cysteine residues, either on the same or on different polypeptide chains. A protein which is made up of two chains is called a *dimer*; proteins comprising three, four and five chains are similarly called *trimers*, *tetramers* and *pentamers* respectively. Where a dimer consists of two identical polypeptide chains this is said to be a *homodimer*. Where the chains are different this is called a *heterodimer*. Haemoglobin, the main protein found in red blood cells, is an example of a tetrameric protein, comprising two identical alpha chains and two identical beta chains.

2.6 Domains

Proteins tend to be folded up into subunits called *domains*. Within a protein a single polypeptide chain can contribute one or more domains. The domains can be structurally similar or quite different from each other. Often a particular property of a protein (for example the ability to bind to another molecule or ligand) can be attributed to one domain of the protein. At the genetic level each domain is likely to be encoded by a separate exon (see Chapter 5).

> **CLINICAL EXAMPLE: DOMAINS IN ANTIBODIES**
>
> Antibodies provide an example of proteins with multiple domains. Antibodies are an important molecular defence against infection. They are multifunctional molecules – firstly, they are able to recognise and bind to other molecules (antigens) with a high degree of specificity; secondly, they can interact with cells and other molecules of the immune system to allow the elimination of an infection. Antibodies are made up of two heavy chains and two light chains. Within each polypeptide chain there are multiple domains, and each domain has a structure comprising two beta sheets. The N-terminal domains are responsible for binding to the antigen, and the domains towards the C-terminus of the heavy chains interact with cellular receptors on phagocytes (see Chapter 25).

2.7 Functions of proteins

Proteins carry out many functions within the body. They may form the structural elements of tissues, important for mechanical support. They can act as transporters to carry other molecules from one location to another. They can also act as hormones, which are chemical messengers that carry information from one part of the body to another. Proteins are present in cell membranes, where they may act as receptors to alert the cell to the presence of molecules in its environment. Proteins also play an important role in defence against infection.

One of the key roles of proteins is to act as *enzymes*. Enzymes are found in both intracellular and extracellular locations within the body and catalyse the various chemical reactions on which life depends. The table overleaf shows examples of the functions carried out by proteins.

2.8 Conformational change

In order to be fully functional, proteins must be folded correctly – they must be in their correct *conformation*. Protein conformation can be disrupted under a number of conditions, including extremes of pH, high temperature and in the presence of detergents. The function of proteins can also be regulated under physiological conditions by altering their conformation. Conformational change can be induced in a number of ways, for example by the binding of a ligand or by covalent modification of protein molecules. *Phosphorylation* is a form of covalent modification of proteins that is commonly used to regulate the activity of enzymes. The process of phosphorylation is itself mediated by enzymes called *protein kinases*. Another way in which protein conformation can be regulated is by *cleavage*. Some enzymes are synthesised as inactive precursors which only become active when they are cleaved. One such enzyme is trypsin which is found in the intestine where it plays a role in breaking down the proteins in food. The enzyme is initially synthesised in the pancreas as an inactive precursor called trypsinogen, which only becomes active when a peptide bond is cleaved. The active enzyme can then catalyse the breakdown of further trypsinogen molecules generating more active enzyme.

Function	Example	Role
Structural	Collagen	Provides tensile strength in connective tissues
Transporter	Haemoglobin	Transports oxygen from the lungs to the various tissues of the body
Hormone	Insulin	Controls the concentration of blood glucose
Receptor	Acetylcholine receptor	Present on muscle cells where it binds acetylcholine leading to muscle contraction
Defence against infection	Antibodies	Bind to an infectious agent and allow its destruction by cells and molecules of the immune system
Enzyme	Trypsin	Breaks down food proteins in the intestine so that they can be absorbed into the blood
Movement	Myosin	Present in muscle where it forms part of the contractile mechanism

2.9 Test yourself

The answers are given on p. 175.

Question 2.1
What are the subunits called that polymerise to form proteins?

Question 2.2
Which two functional groups participate in the formation of a peptide bond?

Question 2.3
What is the primary structure of a protein?

Question 2.4
In an alpha helix
(a) where are the amino acid side chains located?
(b) How many amino acid residues are there per turn of the helix?
(c) What sorts of bonds occur between every fourth peptide bond?

Question 2.5
What is a protein called that comprises: (a) two identical polypeptide chains; (b) two non-identical polypeptide chains?

03 Carbohydrates

> **BASIC CONCEPTS:**
>
> Carbohydrates range in structure from simple sugar molecules, such as glucose, to polymers of sugar molecules, such as those that constitute the rigid wall around all plant cells. Carbohydrates are a major source of energy for the body and so knowledge of their basic structure is very important in understanding energy metabolism. Carbohydrates can be attached to other molecules, including proteins and lipids, and are present on the surface of cells where they can act as recognition molecules.

3.1 Introduction

Carbohydrates, as the name suggests, are molecules made up from carbon and water (hydrogen and oxygen). Carbohydrates serve a number of important functions. They are a major source of energy for life. They are also important structural molecules in many organisms. Additionally, carbohydrates can be attached to other biological molecules, such as proteins, and in doing so modify the properties of the molecule. Probably the most abundant organic molecule on earth is the carbohydrate *cellulose*, a major constituent of plant cell walls. Sugars are small carbohydrates and these can be joined together to form *oligosaccharides* and *polysaccharides*. Different sugars can be joined together in many different ways to give rise to a very diverse array of structures.

3.2 Monosaccharides

The simplest sugars are monosaccharides, which have the general formula $(CH_2O)_n$. According to the value of n, sugars are described as *trioses*, *pentoses*, *hexoses* etc. An important sugar in energy metabolism is *glucose*. This is a hexose with the formula $C_6H_{12}O_6$. Monosaccharides such as glucose can occur as mirror image D and L isomers (*enantiomers*), similar to amino acids (see Fig. 3.1). D or L refers specifically to the arrangement of groups at the asymmetric carbon furthest from the aldehyde group (carbon 5 in glucose).

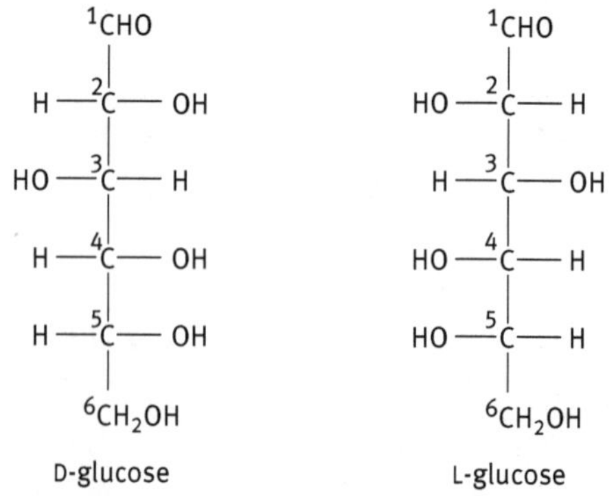

Figure 3.1. D and L glucose

Other sugars, including mannose and galactose, share the same chemical formula as glucose but are not mirror image forms (Fig. 3.2). Glucose frequently assumes a ring form, as shown in Fig. 3.2, by the reaction of the aldehyde group on carbon 1 with the hydroxyl group on carbon 5. The glucose ring can also be found in either alpha or beta form, according to the arrangement of the OH group attached to the first carbon atom. In alpha

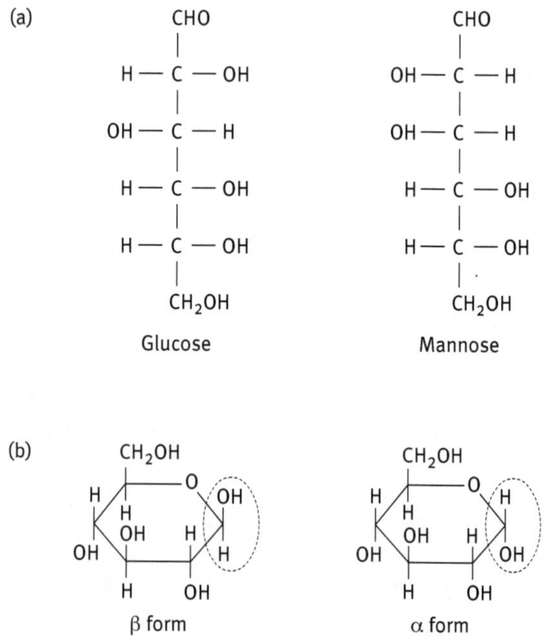

Figure 3.2. (a) Glucose and mannose in open chain form; (b) Glucose in ring form

glucose the OH group is below the plane of the ring, whereas in beta glucose it is above the plane of the ring. This minor structural difference has important consequences for the properties of the molecules and their polymers. Starch is a polymer of alpha glucose molecules and cellulose is a polymer of beta glucose. Note that humans have digestive enzymes that can degrade starch but are unable to digest cellulose.

Sugars may also contain additional groups. Glucosamine is a derivative of glucose in which an amino (NH_2) group replaces one of the OH groups. The amino group frequently has an additional acetyl group ($COCH_3$) attached to it, forming N-acetyl glucosamine. Another important glucose derivative is glucuronic acid, which contains a carboxylic acid group. Sugar derivatives such as N-acetyl-glucosamine and glucuronic acid are present in glycosaminoglycans, which are an important component of the extracellular matrix (see Chapter 15).

3.3 Glycosidic bond

When sugars join together to form oligosaccharides or polysaccharides this creates a *glycosidic bond*. A glycosidic bond is formed by the interaction of an OH group on one sugar with an OH group on another sugar. This involves the elimination of water and is therefore a condensation reaction. Enzymes that can break down glycosidic bonds are termed *glycosidases*. In comparison to the peptide bond, which is always formed between the NH_2 and COOH groups on the alpha carbon of an amino acid, there is much greater diversity in the way in which glycosidic bonds can form. Different carbon atoms can be involved, and an individual sugar can participate in multiple glycosidic bonds to give rise to branched structures. The bond is described according to the particular carbon atoms involved and whether the hydroxyl groups are in the alpha or beta position. In starch the glucose units are joined by α-1,4 linkages (Fig. 3.3), which means that the reaction involves the OH group on carbon 1 of one sugar reacting with the OH group on carbon 4 of the next. Glycosidases are highly specific for a particular type of glycosidic linkage.

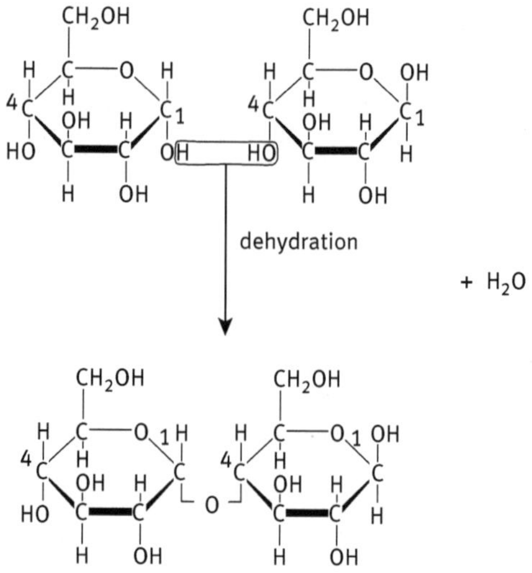

Figure 3.3. Formation of an α-1,4 glycosidic bond

3.4 Polysaccharides

Polysaccharides consist of many monosaccharide subunits covalently linked together. They may involve one or more different sugar units and can be either linear or branched in structure. Starch is a storage polysaccharide of plants that plays an important role in the human diet. Starch has two components – *amylose*, which is a linear polymer of glucose joined by α-1,4 linkages, and *amylopectin*, which is similar but also contains branches due to additional α-1,6 linkages occurring at points along the chain (Fig. 3.4). The enzyme that digests starch is called *amylase* and is present in saliva and in the small intestine.

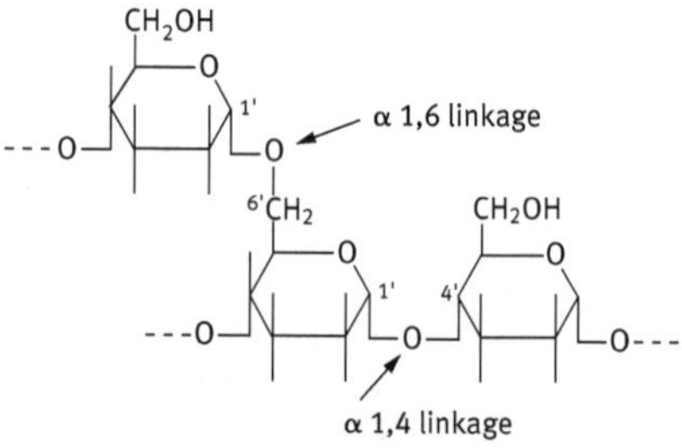

Figure 3.4. α-1,4 and α-1,6 linkages in glycogen

Mammals store carbohydrates in the form of *glycogen* and this can be found in the liver and in muscle. Glycogen is similar in structure to amylopectin but is more branched.

Cellulose is an unbranched polymer of glucose linked by β-1,4 glycosidic bonds and is an important structural component of the cell walls of plants. Humans do not have an enzyme that can hydrolyse this type of linkage and cannot use cellulose as an energy source. However, cellulose does play an important role in digestive function as a source of roughage, or insoluble matter, helping to maintain the consistency of the faeces and to stimulate mucus secretion as waste matter passes through the large intestine. Ruminants, such as cattle and sheep, are able to make use of cellulose because they harbour bacteria that secrete the enzyme cellulase in a part of their digestive tract known as the rumen.

3.5 Glycoconjugates

Other molecules may also have sugars attached to them – a process known as *glycosylation*. Many proteins are glycosylated, particularly those that are found in cell membranes or that are secreted from the cell. Glycoproteins contain one or more oligosaccharides attached to the protein backbone, either at the hydroxyl group of serine or threonine residues (when they are described as O-linked) or via the amide group of asparagine (N-linked). The relative amount of carbohydrate in glycoproteins varies enormously, from merely 1 or 2% to over 70%. Mucin is a very large and heavily glycosylated molecule that forms a major constituent of mucus and contributes to its gel-like properties. Mucin contains a very high proportion of carbohydrate in the form of O-linked oligosaccharides. Lipids may also be glycosylated.

CLINICAL EXAMPLE: ABO BLOOD GROUPS

The human ABO blood group antigens are oligosaccharides present on glycoproteins and glycolipids on red blood cells. Blood group specificity is defined by the oligosaccharide component. Enzymes called glycosyl transferases are required for the synthesis of oligosaccharides. These are highly specific and whether an individual produces the blood group oligosaccharide O, A or B will depend on which glycosyl transferases they have inherited from their parents. If an individual with type O blood were to receive a transfusion of type A blood then their immune system would react against the A oligosaccharide, destroying the transfused cells and leading to serious consequences.

3.6 Functions of carbohydrates

Carbohydrates serve a range of biological functions, several of which have been referred to above. They are an important part of the diet, providing a ready source of energy. In the form of polysaccharides such as starch and glycogen they play a role in storage. They are also important structural components of cells (e.g. cellulose) and influence the mechanical properties of tissues or secretions (e.g. glycosaminoglycans, mucins).

Sugars are also important *recognition molecules*. Within the cell, the carbohydrates on glycoproteins play a role in ensuring that these molecules are transferred to the appropriate cellular compartments. Carbohydrates present on cell surface glycoproteins and glycolipids also interact with proteins called *lectins*. This type of interaction plays an important role in the trafficking of white blood cells to sites of infection (see Chapter 25). Oligosaccharides also play an important role in the recognition of an egg by a sperm (see Chapter 11).

3.7 Test yourself

The answers are given on p. 175.

Question 3.1
What is the chemical formula for glucose?

Question 3.2
What is the name of the bond that joins sugars together in a polysaccharide?

Question 3.3
What is the enzyme that degrades starch called?

Question 3.4
What are the enzymes that synthesise oligosaccharides?

Question 3.5
What polymer is used to store carbohydrates in animals?

⬡04 Lipids

BASIC CONCEPTS:
Lipids are commonly called fats or waxes. They are generally insoluble in water. They include phospholipids, which have a hydrophilic head region and a hydrophobic tail. Lipids are an important constituent of cell membranes where their properties determine membrane permeability. Lipids can also act as longterm energy stores in the body.

4.1 Introduction

Lipids are molecules of varying sizes and structures that tend to be insoluble in water. Lipids are a major constituent of biological membranes, where they form *bilayers*. Lipids are also an important source of energy yielding approximately twice the energy value of carbohydrates on a weight-for-weight basis.

4.2 Fatty acids

Fatty acids consist of a long hydrocarbon chain with a terminal carboxyl (COOH) group. Fatty acids vary in the length of their carbon chain and in the degree of saturation – that is, how many hydrogen atoms are bound to the various carbon atoms (see Fig. 4.1 for examples). Fatty acids with the maximum number of hydrogen atoms are described as *saturated*. *Unsaturated* fatty acids have fewer hydrogen atoms attached to the carbon chain and contain double bonds between two or more of the carbon atoms. Those that have multiple double bonds are *polyunsaturated*. The degree of saturation affects the physical properties, with unsaturated fats tending to have a lower melting point than saturated fats. Polyunsaturated fats in the diet are believed to be beneficial because they help to lower the levels of cholesterol in the blood and reduce the risk of coronary heart disease.

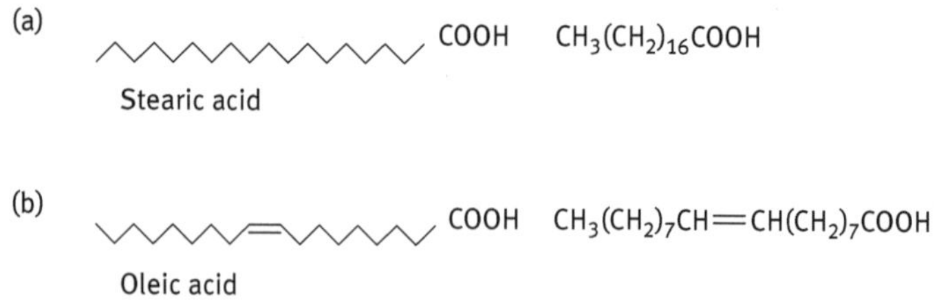

(a)

Stearic acid

COOH $CH_3(CH_2)_{16}COOH$

(b)

Oleic acid

COOH $CH_3(CH_2)_7CH=CH(CH_2)_7COOH$

Figure 4.1. (a) Stearic acid – a saturated fatty acid found in animal fat; (b) Oleic acid – an unsaturated fatty acid found in olive oil

4.3 Triglycerides and phospholipids

The carboxylic acid group on a fatty acid can react with a hydroxyl group on the three-carbon alcohol glycerol to form a *glyceride*. Triglycerides (Fig. 4.2) consist of three fatty acid molecules attached to one glycerol molecule. This is the main form in which fat is stored in the body and provides a highly efficient energy reserve.

In *phosphoglycerides*, which are a major class of lipid in cell membranes, two of the OH groups of glycerol have a fatty acid molecule attached, while the other OH is attached via a phosphate group to a polar head group. The head group varies and gives phospholipids their names. Commonly occurring phospholipids include phosphatidyl serine, phosphatidyl choline and phosphatidyl inositol, in which the head groups are serine, choline and inositol respectively. Another type of phospholipid is based not on glycerol but on the molecule sphingosine.

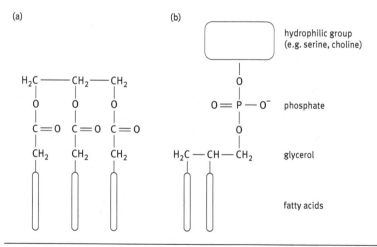

Figure 4.2. General structure of a triglyceride (a) and a phospholipid (b)

The fatty acid chains of phospholipids (Fig. 4.2) are hydrophobic, whereas the polar head groups are hydrophilic, making phospholipids *amphipathic*. In an aqueous environment these molecules will tend to arrange themselves so that the hydrophobic fatty acid chains interact with each other and the head groups are exposed to water. These interactions give rise to the phospholipid bilayer that is characteristic of cell membranes (see Chapter 6 and Fig. 4.3).

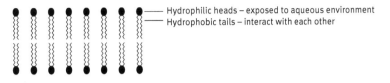

—— Hydrophilic heads – exposed to aqueous environment
—— Hydrophobic tails – interact with each other

Figure 4.3. A lipid bilayer

4.4 Cholesterol

Another important lipid found in mammalian cell membranes is *cholesterol*. Cholesterol has a very different structure from the fatty acid type lipids described above. Instead of having long hydrocarbon chains, cholesterol is based on a structure containing four linked hydrocarbon rings with a short hydrocarbon chain. Cholesterol is found in varying proportions in different membranes and plays an important role in determining membrane properties such as fluidity and permeability.

> **CLINICAL EXAMPLE: FAMILIAL HYPERCHOLESTEROLAEMIA**
>
> Cholesterol absorbed from the diet circulates in the bloodstream bound to carrier proteins in a complex known as a low density lipoprotein (LDL). It is taken into cells after binding to a specific LDL receptor on the cell surface. An inherited defect in the LDL receptor can lead to failure to internalise cholesterol with a consequent raised level of cholesterol in the blood. Because of their raised blood cholesterol levels, affected individuals will have a greater likelihood of developing heart attacks and strokes. In individuals with two copies of the defective LDL receptor gene, circulating cholesterol is so high that they are likely to die in their early thirties.

4.5 Functions of lipids

The discussion so far has focused on lipids as a primary constituent of cell membranes. Lipids have other important biological functions. Triglycerides can be stored in *adipose tissue* (fat) to provide a reserve of energy. Fatty tissues also play a role in thermal insulation. Lipids are an important component of the myelin sheath, which surrounds neurons and provides electrical insulation, allowing the effective transmission of nerve impulses.

4.6 Test yourself

The answers are given on p. 175.

Question 4.1
What is the term used to describe fatty acids in which the maximum number of hydrogen atoms are attached to the carbon chain?

Question 4.2
What is the name of the three-carbon alcohol that occurs in phospholipids such as phosphatidyl choline?

Question 4.3
How many fatty acid chains are present in a molecule such as phosphatidyl choline?

Question 4.4
What is the term used to describe molecules, such as phospholipids, which have a hydrophilic and a hydrophobic region?

Question 4.5
What is the name of the tissue used to store fat in mammals?

05 Nucleic acids and genes

BASIC CONCEPTS:

DNA is the basic molecule of life. A DNA molecule consists of two strands of nucleic acid which are capable of replication. A chromosome consists of a long strand of DNA with associated proteins and each human cell contains 46 chromosomes. Each chromosome carries a series of genes, which are sections of DNA in which the nucleic acid sequence codes for the production of a specific protein. The genetic code is based on triplets of nucleotides, which specify which amino acid goes into a particular position in a protein molecule. The structure and function of an individual cell is determined by the genes which are active in that cell. An understanding of how genes are turned on and off in specific cells is necessary to appreciate the diversity found within individuals.

5.1 Introduction

Nucleic acids hold the key to life. They are the stuff that *genes* are made of – the macromolecules containing the information that defines species and individuals. There are two main nucleic acids: *DNA*, or deoxyribonucleic acid, and *RNA*, or ribonucleic acid. In most organisms, including humans, genetic information is encoded within DNA but is then copied (transcribed) into RNA before being translated into proteins.

Nucleic acids (see Fig. 5.1), like proteins and polysaccharides, are polymers made up of a large number of subunits joined together. The subunits of DNA and RNA are called *nucleotides*. DNA and RNA are synthesised by copying from existing DNA or RNA templates using enzymes called *polymerases*. In order for a cell to divide it must first undergo DNA synthesis to produce a new copy of the genetic material. In order to produce proteins, cells must first transcribe the genes in the DNA into RNA.

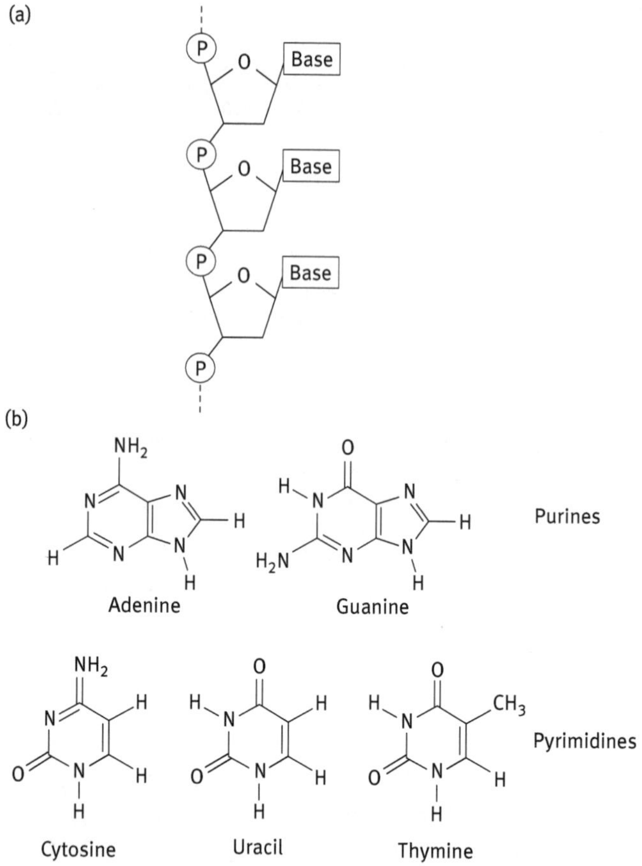

(a)

(b)

Adenine   Guanine   Purines

Cytosine   Uracil   Thymine   Pyrimidines

Figure 5.1. (a) Basic structure of a nucleic acid; (b) Structure of the five bases present in DNA and RNA

5.2 DNA and RNA

The nucleotides that make up both DNA and RNA comprise three parts – a five-carbon sugar (ribose in RNA, deoxyribose in DNA), a phosphate group and an organic base. Four different bases are used in DNA – adenine, thymine, guanine and cytosine. In RNA, uracil replaces thymine. Adenine and guanine contain two joined rings and are called *purines*; cytosine, thymine and uracil contain a single ring and are called *pyrimidines*. The backbone of the nucleic acid is formed by the sugars joined through phosphate groups. The OH on the third carbon of one sugar is covalently bound to the phosphate group, which in turn forms a bond with the OH on the fifth carbon of the next sugar. This is called a 3′–5′ *phosphodiester* bond. At one end of the chain a phosphate group is attached to the 5′-carbon of the sugar, and this is called the 5′-end – the other end has a free 3′-OH group and this is called the 3′-end.

The bases protrude from the sugar–phosphate backbone of the molecule, and the key to the function of DNA and RNA is the ability of certain bases to interact non-covalently, by means of hydrogen bonds, to form *base pairs*. Adenine has the ability to pair with thymine (or uracil in RNA) and guanine has the ability to pair with cytosine. The DNA molecule is double-stranded and the two strands are complementary. Where adenine occurs on one strand of the molecule thymine is found on the other strand where it can form a base pair. Similarly guanine occurs in the opposite position to cytosine. This allows for the two strands of DNA to interact in a highly regular fashion with the sugar–phosphate backbone on the outside of the molecule and the paired bases on the inside. The double-stranded DNA forms a double helix (Fig. 5.2), with ten bases per turn of the helix. The base pairing only works if the two chains run in opposite directions – one chain lies 5′–3′ and the other lies 3′–5′.

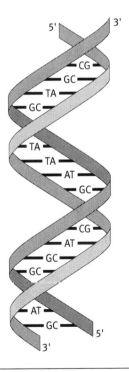

Figure 5.2. The DNA double helix showing base pairing between strands

RNA is a single-stranded molecule, though it may fold back on itself in such a way as to allow internal pairing of some of the bases. There are three main types of RNA within a cell – *messenger* RNA (mRNA), *transfer* RNA (tRNA) and *ribosomal* RNA (rRNA). Each of these has a particular role in protein synthesis.

5.3 DNA synthesis

In order for cells to divide they must first reproduce their DNA. DNA is synthesised in a process called *semi-conservative replication*. The two strands of the double helix become separated from each other, and then each acts as a template for the synthesis of a new and complementary strand. When synthesis is complete, each daughter cell inherits DNA in which one strand is derived from the original DNA and the other is newly synthesised. Accuracy is crucially important in DNA synthesis and so mechanisms exist to correct any errors that might arise. Several enzymes are involved in DNA synthesis – *helicases* separate the two strands prior to replication; *polymerases* allow extension of the new strand by the addition of nucleotides. A third type of enzyme, called a DNA *ligase*, is also required. This is because DNA synthesis proceeds continuously on only one of the two strands – on the other strand DNA is made in small fragments which must subsequently be joined together. This is the function of the ligase.

5.4 RNA synthesis

Like DNA, RNA is also synthesised by polymerases in a template-dependent fashion. DNA forms the template for RNA synthesis. Synthesis of RNA is initiated at sites within the DNA known as *promoters* – these are characterised by particular sequences of bases found before the start of genes. The promoters of most genes have a characteristic sequence known as the TATA box located about 30 bases upstream of the site of initiation of RNA synthesis. The TATA box is so called because of its sequence (typically TATAAA). As with DNA, RNA synthesis also requires the action of helicases to open up the DNA so that the polymerase can gain access to the gene. In addition to the RNA polymerase, other proteins, known as *basal transcription factors*, are required for transcription to be initiated in eukaryotes. The TATA box plays a key role in allowing the assembly of the various proteins required for transcription of DNA by the RNA polymerase. RNA is then produced by the sequential addition of nucleotides complementary to those in the DNA, guided by the rules of base pairing. An important difference between DNA synthesis and RNA synthesis is that in the latter case only one of the two strands of DNA is copied.

5.5 The genetic code

It had been established by the middle of the twentieth century that the genetic information in an organism was contained in DNA. Further experiments established that genes encoded proteins, giving rise to the 'one-gene one-polypeptide hypothesis'. The discovery of the genetic code provided the vital link that connected DNA and proteins.

Central to the genetic code is the existence of *codons*. A codon is a sequence of three bases that specify a particular amino acid. Each base is part of only one codon, and so the code is said to be *non-overlapping*. A single codon can specify only one amino acid, and the same codon always encodes the same amino acid, regardless of the species, so the code is said to be *universal* (although it should be noted that there are some exceptions to this rule, for example in mitochondrial DNA). As there are four different bases this gives rise to a possible 64 ($4 \times 4 \times 4$) codons. There are only 20 amino acids and for most of these there is more than one codon, so the code is said to be *degenerate*. For example, the amino acid tyrosine is coded for by the codons UAC and UAU. Some codons (UAA, UAG, UGA) do not specify an amino acid but instead provide a signal to terminate protein synthesis – these are called *stop codons*. There is also a *start codon* (AUG, which codes for methionine) which provides the signal to allow translation to be initiated.

5.6 Protein synthesis

For proteins to be synthesised, a DNA gene must be transcribed into a messenger RNA which is then translated into protein. This flow of information from DNA $\rightarrow$ RNA $\rightarrow$ protein is sometimes called the *central dogma* (Fig. 5.3).

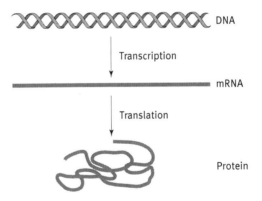

Figure 5.3. Flow of information from DNA to RNA to a protein

CLINICAL EXAMPLE: HIV BREAKS THE CENTRAL DOGMA

When the central dogma was proposed it was believed that the flow of information could occur in only one direction – from DNA to RNA to protein. Subsequently, viruses were discovered that were able to copy RNA into DNA by means of an enzyme called *reverse transcriptase*. These are called *retroviruses* and include the human immunodeficiency virus HIV, which is responsible for AIDS. The genetic information in HIV is contained within RNA. In infected cells this is then copied into DNA under the action of reverse transcriptase. The double-stranded DNA that is produced can then insert itself into the DNA of the host cell. The DNA copy of the viral RNA can subsequently be transcribed to produce mRNA which is then translated to produce new viral proteins. Because reverse transcriptase is specific to the virus and not required for the functioning of the host cells, reverse transcriptase has been the target of some of the antiviral drugs that have been developed to treat HIV infection.

Transcription of genes proceeds as described above. Once mRNA has been produced this then becomes associated with the *ribosomes*, which are the site of protein synthesis. Ribosomes are particles made up of protein and RNA and are responsible for catalysing the formation of peptide bonds between amino acids as proteins are synthesised.

As protein synthesis proceeds, the ribosome moves along the mRNA. Amino acids, the precursors of proteins, are picked up by adaptor molecules of tRNA. For each amino acid there is a particular tRNA that is able to bind only that amino acid. The tRNA also contains a site known as the *anticodon*, which comprises three bases complementary to the codon for that amino acid. This allows tRNA molecules to align themselves with codons within the mRNA so that this is translated into the correct sequence of amino acids (Fig. 5.4).

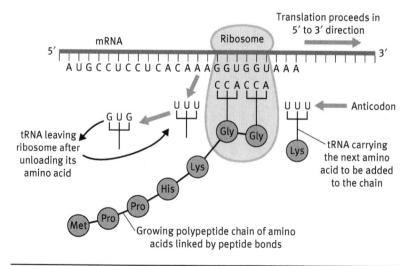

Figure 5.4. Translation of mRNA into protein

Amino acids are added sequentially to a growing polypeptide chain under the influence of the catalytic activity of the ribosome. Interestingly, this activity seems to be associated not with ribosomal protein enzymes but with rRNA, giving rise to the concept of a *ribozyme*, or catalytic RNA. This *peptidyl transferase* activity allows the formation of peptide bonds between the NH_2 group of an incoming amino acid on a tRNA molecule and the free COOH group of the preceding amino acid. After formation of the peptide bond the tRNA can be released and reused.

5.7 Introns and exons

In mammals, and other higher organisms, genes do not consist of continuous sequences of DNA but of coding sequences which are expressed (*exons*) separated by non-coding intervening sequences (*introns*). Both the exons and introns of a gene are transcribed into RNA and then the introns are removed. Within proteins an individual domain may be encoded by an exon.

5.8 Regulation of gene expression

Cells need to regulate the expression of their genes for various reasons. Some proteins need to be expressed under particular environmental conditions, and it would be a waste of valuable resources to produce these when they are not needed. Additionally, in complex multicellular organisms, different cells perform specialised functions and so the repertoire of proteins they express will vary accordingly. Within any particular cell some genes will be expressed *constitutively* (all the time), regardless of the environment, and some will be expressed *inducibly* when their protein products are needed.

Regulation of gene expression involves a number of mechanisms. In bacteria, genes encoding related proteins in a pathway are clustered together into *operons*. Regulation of expression of the genes within an operon can be influenced by proteins that bind to the DNA and affect accessibility to the RNA polymerase. The classic example of this is the *lac* operon, described by Jacob and Monod in 1960. The lac operon encodes genes that enable cells to use the substrate lactose as an energy source. In the absence of lactose the genes are not expressed because a *repressor* protein is bound to a region of the operon called the *operator*, where it prevents the RNA polymerase from binding. In the presence of lactose, an *inducer* binds to the repressor molecule, reducing its affinity for the operator. The inducer is not lactose itself but allolactose, an isomer produced from lactose.

Regulation of gene expression in eukaryotes is more complex. DNA is present in chromosomes where it is bound tightly to proteins known as *histones*. Histones are an important structural component of chromosomes and play a role in the regulation of gene expression by regulating the accessibility of promoters. In this way they can determine whether or not transcription may begin. Histones are clustered in groups of eight (octamers) and each octamer is entwined in DNA to form a *nucleosome*. Histones can be covalently modified in such a way as to alter their affinity for DNA. Acetylation of histones leads to reduced affinity for DNA so that it becomes more accessible.

The importance of the TATA box in initiating transcription has been noted above. Additionally there may be other sites upstream of this that can bind transcription factors and enhance or repress expression of the gene. Sequences found within DNA called *enhancers* also play an important role in regulating transcription. Enhancers can be found at a distance of 100s–1000s of base pairs from a gene that they influence, but can interact with the relevant promoter region by the looping of DNA. Enhancers bind to specific regulatory proteins which facilitate transcription. The production of regulatory proteins able to bind to particular enhancers is specific to certain types of differentiated cells and this ensures that cells make the range of proteins appropriate to their particular function.

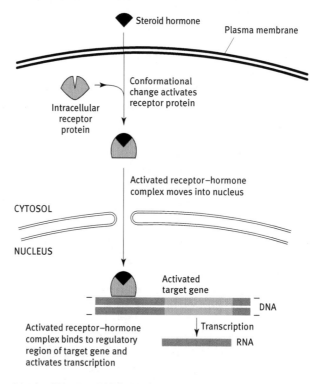

Reproduced from *Essential Cell Biology* with permission from Garland.

Figure 5.5. Regulation of gene expression by a steroid hormone

A simple example of the regulation of gene expression in mammals is provided by the response that is seen to steroid hormones (see Chapter 18). These molecules are hydrophobic and so they pass through cell membranes and bind to intracellular receptor molecules. These molecules, once the hormone has bound, are then able to move into the nucleus where they bind to special sites in the DNA to activate transcription of selected genes (Fig. 5.5).

CLINICAL EXAMPLE: ENHANCERS, ONCOGENES AND CANCER

Cells of the specific immune system called B lymphocytes are responsible for the production of immunoglobulins (antibodies), which are important proteins in defence against infection. B lymphocytes are the only cells able to synthesise immunoglobulins, and the production of these proteins is under the regulation of an immunoglobulin enhancer. In some types of cancer affecting B cells, a cellular gene called myc is brought into the vicinity of the immunoglobulin enhancer following a chromosomal translocation, when a piece of the chromosome containing the myc gene is moved to the chromosome containing the immunoglobulin genes. The resulting over-expression of myc leads to unregulated proliferation of the cells giving rise to cancer.

5.9   Test yourself

The answers are given on p. 175.

Question 5.1
What does the abbreviation DNA stand for?

Question 5.2
What are the four bases found in DNA? Which pairs with which?

Question 5.3
What is a sequence of three bases that specifies an amino acid called?

Question 5.4
What term is used to describe the DNA sequence at which RNA synthesis is initiated?

Question 5.5
What are the molecules that bind amino acids and match them to codons in mRNA?

06 The cell

> **BASIC CONCEPTS:**
> The cell is the basic structural unit of life. In eukaryotic cells the genetic material is contained within the nucleus. Multicellular organisms consist of assemblies of specialised cells organised into tissues and organs. Each cell is surrounded by a plasma membrane which regulates its interactions with the external environment. Within each cell, basic functions such as respiration, protein synthesis and excretion are compartmentalised into internal structures known as organelles. Knowledge of the functions of these organelles is a prerequisite for understanding the most basic concepts of cell biology.

6.1 Introduction

All living organisms are made of *cells*. Cells are membrane-bound units in which many of the reactions essential to life occur. The simplest organisms consist of a single cell and are sometimes referred to as *unicellular*. Single-celled organisms may be described as either *prokaryotic* or *eukaryotic*. Eukaryotic cells are characterised by having a nucleus and have a more complex cellular architecture than prokaryotic cells. More complex, *multicellular* organisms including plants, animals and fungi are also eukaryotes. Bacteria are prokaryotes and their cells lack a nucleus and tend to be smaller than eukaryotic cells.

6.2 Eukaryotic cells

Membranes

All cells are surrounded by a plasma *membrane* (Fig. 6.1). Membranes consist of a phospholipid bilayer. Phospholipids are amphipathic molecules, with a hydrophilic head group and hydrophobic fatty acid chains (see Chapter 4). These are arranged so that the hydrophilic head groups are exposed on the surface (either external or internal to the cell) and the fatty acid chains are shielded from the aqueous environment and interact with each other.

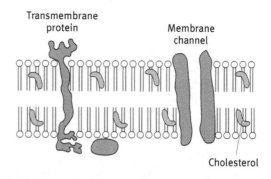

Figure 6.1. The fluid mosaic model of the cell membrane

In addition to phospholipids, membranes also contain other lipids such as cholesterol, and proteins which can contribute up to 45% of the mass of a membrane. Membranes are sometimes described as a fluid mosaic (Fig. 6.1) because the proteins and lipids can freely diffuse in the plane of the membrane.

In eukaryotic cells there are subcellular structures called *organelles* (Fig. 6.2), which may also be surrounded by membranes. Different organelles have different functions within the cell and allow reactions to take place in separate compartments. Some reactions take place on membranes and are catalysed by enzymes that are membrane-bound.

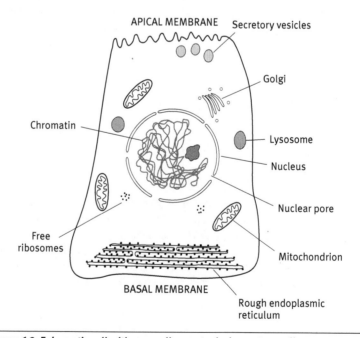

Figure 6.2. Eukaryotic cell with organelles – a typical secretory cell

The nucleus

The nucleus is the organelle that contains the genetic material in the form of *chromosomes*. Chromosomes consist of DNA and associated proteins. The nucleus is surrounded by a double membrane, which separates it from the cytoplasm of the cell. The nuclear membrane contains pores that allow material to move between the nucleus and cytoplasm. It is within the nucleus that DNA and RNA synthesis take place. Following transcription, mRNA leaves the nucleus via the pores in the nuclear membrane and enters the cytoplasm, where it becomes associated with ribosomes and is translated into protein.

Cytoplasm

The part of the cell that is not the nucleus is referred to as the *cytoplasm*. Many of the reactions of life occur in the cytoplasm. The cytoplasm is the principal site of protein synthesis (see Chapter 5) and this is carried out by ribosomes. Ribosomes may be found either free in the cytoplasm or attached to membranes of the endoplasmic reticulum.

Endoplasmic reticulum

The *endoplasmic reticulum* (ER) consists of a collection of interconnected membrane-bound sacs. These are the sites of production of new membrane, and of proteins destined for secretion from the cell. ER that has ribosomes attached to it is described as rough endoplasmic reticulum. Proteins are synthesised by ribosomes attached to the cytoplasmic face of the ER membrane and then transported into the lumen of the ER. Within the ER, proteins have sugars added to them – this is described as a post-translational modification – and they also become correctly folded.

ER which is not specialised for protein synthesis is known as smooth endoplasmic reticulum. This can serve a variety of functions including calcium storage, synthesis of steroid hormones and detoxification.

Golgi apparatus

The *Golgi apparatus* is a series of membrane-bound sacs which are in a constant state of flux as vesicles arrive, depart and move between various zones of the Golgi. After synthesis in the ER, proteins destined for secretion are transported to the Golgi where they undergo further post-translational modification. They are then sorted and transported to the appropriate destination, which may be the cell surface, a lysosome or a secretory vesicle. Membrane lipids are similarly transported through the Golgi on their way to being incorporated into the cell membrane.

Lysosomes

Lysosomes are membrane-bound vesicles that contain a range of hydrolytic enzymes, including proteases. The main function of lysosomes in a cell is to provide a site of intracellular digestion. Within phagocytic cells of the immune system (see Chapter 25) lysosomes play an important role in the destruction of infectious agents that have been taken into the cell.

Mitochondria

Mitochondria (singular mitochondrion) constitute a complex system of organelles found within the cytoplasm of the cell. Like the nucleus, mitochondria are also surrounded by a double membrane, and the inner mitochondrial membrane is extensively folded. The mitochondrial system within the cell is responsible for the production of energy in a process called oxidative phosphorylation (see Chapter 8). Mitochondria contain their own DNA, which encodes for some mitochondrial proteins. They also contain the enzymes needed for oxidative phosphorylation to occur.

CLINICAL EXAMPLE: MITOCHONDRIAL MYOPATHY

Mitochondrial myopathy is a condition in which an inherited defect in one of the proteins involved in oxidative phosphorylation leads to a gradual loss of muscle function. It is thought to affect muscle cells in particular because of their high energy demands. The affected genes are usually in the mitochondrial DNA and so it is a condition that is passed from a mother to her children. Only the daughters of an affected individual can then pass it on to their children. The muscle damage is often seen first in the muscles around the hips and sufferers will complain of muscle pain and difficulty in getting up from a seated position.

The cytoskeleton

Within the cell is a framework of protein filaments that make up the cytoskeleton. The cytoskeleton is responsible for maintaining a cell's shape, mechanical strength and directing the movement of organelles. It also plays an important role in the process of cell division by mitosis (see Chapter 10). In phagocytic cells that can ingest material from the surroundings, the cytoskeleton is also important in the process of phagocytosis.

The cytoskeleton is made up of three types of filaments. Microfilaments consist primarily of the protein actin, which is also found in muscle, and are found in highest density just beneath the plasma membrane. Intermediate filaments are made up of multiple rod-like proteins and extend across the cytoplasm providing mechanical strength. Microtubules consist of the protein tubulin, and these play an important role in mitosis and internal transport through the cytoplasm.

The cytoskeleton is a dynamic structure, in which the filaments assemble and disassemble as required.

6.3 Cell specialisation

Although all eukaryotic cells have the same basic components they can become specialised (differentiated) to allow them to perform their particular functions. For example, red blood cells do not have a nucleus but are filled with the protein haemoglobin, which allows them to transport oxygen to tissues. Nerve cells have a cell body that contains the nucleus and an extended projection, called an axon, along which nerve impulses are conveyed. Cells whose main function is to secrete proteins may have a very well-developed rough endoplasmic reticulum, as in the case of plasma cells that secrete antibody molecules. Sperm cells have a long tail or flagellum that allows them to swim up the reproductive tract – the tail contains abundant mitochondria to provide the necessary energy. Fat cells are specialised for storage of fat, and contain a large lipid globule surrounded by only a thin rim of cytoplasm. Cells in the gut, which function to absorb nutrients, have a greatly increased surface area due to the presence of many short finger-like projections called *microvilli*. Cells in the airways have hair-like protrusions, called *cilia*, which allow them to move a stream of mucus along the surface of the airway.

Plant cell specialisations

Unlike animal cells, plant cells are surrounded by a rigid cell wall made up of the polysaccharide cellulose. Plant cells also have a special type of organelle, the *chloroplast*, containing the green pigment chlorophyll. Chloroplasts are essential for the process of photosynthesis in which plants convert carbon dioxide and water into glucose using light as a source of energy.

6.4 Test yourself

The answers are given on p. 175.

Question 6.1
Which membrane-bound organelles contain hydrolytic enzymes?

Question 6.2
Where in the cell is energy mainly produced?

Question 6.3
In animal cells, apart from the nucleus, which membrane-bound organelles contain DNA?

Question 6.4
What protein occurs in microfilaments?

Question 6.5
Which cytoplasmic particles, found in both prokaryotic and eukaryotic cells, play a central role in protein synthesis?

07 Microorganisms

> **BASIC CONCEPTS:**
> The microbial world comprises a diverse range of organisms, which differ in their structure and biology. These include:
>
> - bacteria, which are prokaryotes
> - fungi and protozoa, which are eukaryotic organisms
> - viruses, which lack the cellular machinery for replication and so must infect cells in order to reproduce.
>
> The human body plays host to a large number of microbial species, most of which are harmless. However, a small number of species can cause infectious disease in animals or humans and these are a major cause of death, particularly in some parts of the world. Immunisation can protect against some microbial diseases and specific classes of drugs are available to treat bacterial, fungal, viral and other types of infection. Some microbes, however, have evolved to develop resistance to current therapies and so there is an urgent need to develop new treatments and vaccines for the future.

7.1 Introduction

Microorganisms, or microbes as they are sometimes known, are defined by their size. Often unicellular, they cannot be seen with the naked eye. Microorganisms are extremely important in the medical sciences, principally in relation to the diseases that some of them can cause. However, it is important to note that most microbes are harmless and many are beneficial to humankind, playing a central role in, for example, the cycling of nutrients, the production of foods and beverages including bread, beer and wine, and the production of drugs such as antibiotics. Those organisms that can cause disease are referred to as *pathogens*.

Pathogens can be transmitted in various ways. Some, such as those that cause the common cold and influenza, infect the respiratory tract and can be transmitted from person to person via droplets produced by coughs and sneezes. Other infections, such as *Salmonella* and cholera, are acquired by eating or drinking contaminated food and water. Some infections may be sexually transmitted, such as HIV and syphilis. In other cases, microorganisms may enter the body through cuts or other damage to the skin: this is the case for tetanus, which is caused by a bacterium that is found in soil. Some infectious agents, such as the one that causes malaria, are spread by insects when they bite humans or other animals. In this case the insect is described as a *vector* for the disease. The table below shows the major causes of death due to infectious diseases.

Disease	Causative agent	Deaths in 2009/10
Total deaths due to infectious and parasitic diseases	Various	6.7 million
HIV–AIDS	RNA virus	1.8 million
Tuberculosis	Bacterium	1.4 million
Malaria	Single celled eukaryotic organism, *Plasmodium*	0.7 million
Measles	RNA virus	0.2 million

Data from WHO World Health Report 2012.

7.2 Bacteria

All plants, animals and fungi are eukaryotes (see Chapter 6), whereas bacteria (singular: bacterium) are prokaryotes. Prokaryotic cells (see Fig. 7.1) are smaller than eukaryotic cells and by definition do not have a nucleus. Their DNA consists of a single circular strand, which lies free in the cytoplasm. Prokaryotes lack most of the membrane-bound organelles described for eukaryotic cells. They do have ribosomes, which are smaller than those of eukaryotes, and these are found only in free form because there is no endoplasmic reticulum. Bacteria, like plant cells, are surrounded by a rigid cell wall (see Section 6.3), and in bacteria this is made up of a substance called *peptidoglycan*, which consists of polysaccharide chains cross-linked by short peptides. Bacteria can be differentiated using Gram stain into two main groups according to the structure of the cell wall:

- Gram positive bacteria have a thick layer of peptidoglycan and are stained dark blue in Gram stain
- Gram negative bacteria have a much thinner layer of peptidoglycan and have an additional outer membrane that lies outside this; they appear pink in Gram stain.

Some bacteria are further enclosed in a polysaccharide capsule, and some have additional structures extending from the cell including:

- *flagella* (singular: flagellum), which are involved in movement
- *pili* (singular: pilus), which are involved in the transfer of genetic material between bacterial cells in a process called conjugation; this is of medical relevance because it can allow the transfer of DNA encoding for antibiotic resistance between bacteria.

Bacteria are often described according to their shape: cocci (singular: coccus) are spherical, whilst bacilli (singular: bacillus) are rod-shaped.

Bacteria of medical importance include both Gram positive and Gram negative species. Diseases caused by bacteria include tuberculosis (TB), leprosy, and food poisoning due to *Salmonella*.

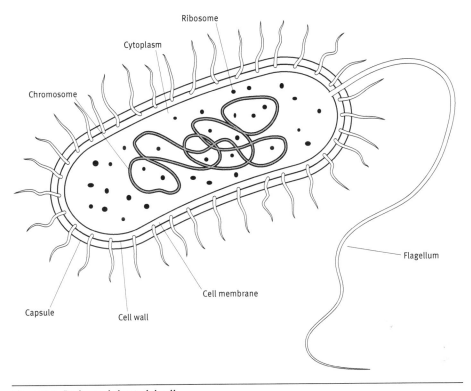

Figure 7.1. Prokaryotic bacterial cell

Bacteria undergo cell division by *binary fission*, a process of asexual reproduction. The bacterial DNA replicates to give rise to two identical copies (see Section 11.1). The cell then divides to produce two daughter cells, each with a single copy of the DNA. This process can be very rapid, with a generation time of as little as 10 minutes. The ability of bacteria to increase their numbers in a short space of time is highly significant in infectious disease, when the early stages of infection can be viewed as a race between the microbe to establish itself and the immune system that is trying to protect the body from the infection (see Chapter 25).

Some bacteria have the ability to form spores, and as spores they can persist for many years in adverse environmental conditions. During the Second World War scientists were interested in the potential use of anthrax as a biological weapon. As an experiment they detonated bombs containing anthrax spores on Gruinard, a small island off the west coast of Scotland. Sheep that were present on the island soon contracted the disease and died. The island remained infected with anthrax for more than four decades, showing the length of time spores can survive, until it was finally decontaminated in the 1980s.

7.3 Viruses

Viruses are organisms that cannot survive independently: they can only reproduce themselves by infecting a host cell and making use of its genetic machinery for replication. Viruses can infect plant or animal cells. Some can even infect bacteria and these are called bacteriophages. Human diseases caused by viruses include HIV/AIDS (human immunodeficiency virus / acquired immunodeficiency syndrome), chickenpox, measles and influenza ('flu' for short).

Viruses are very small, typically ranging between approximately 5 and 300 nm (1 nm is 10^{-9} m). They consist of a genome that can be either single-stranded or double-stranded RNA or DNA. The genome can be circular or linear, or in some cases it can be divided up into multiple strands (segmented). The viral genome is surrounded by an outer protein coat or capsid and some viruses are further enclosed by a lipid envelope. On their surface viruses often have proteins that are involved in entry into a cell. The influenza virus has a protein called haemagglutinin on its surface that binds to sialic acid molecules on respiratory epithelial cells, initiating entry of the virus into the cell. Another surface protein of the influenza virus, neuraminidase, is involved in the release of newly synthesized virus particles from an infected cell. Virus particles may also contain enzymes required for their replication.

Several steps are involved in viral replication. First, the virus must bind to a cell and gain entry. This often involves endocytosis (see Section 9.5). Once inside the cell the viral genome is released from the virus particle (a process known as uncoating) undergoes replication, and viral proteins are synthesized using the host cell machinery. Finally, new virus particles assemble and are released from the cell, a process which can involve cell lysis or, in the case of enveloped viruses, budding from the membrane of the cell (Fig. 7.2).

Viruses display some interesting genetic features. Many have genes that overlap, so that the same sequence of nucleotides forms part of the coding sequence for different proteins. For example, the hepatitis B virus, a DNA virus that causes liver disease, has a genome encoding a number of proteins

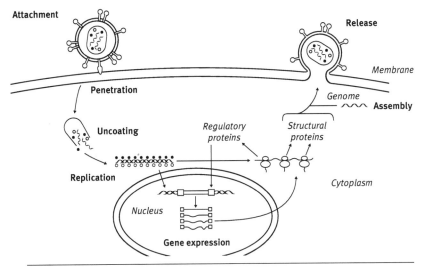

Figure 7.2. Stages in viral replication

including a polymerase that is involved in replication of the virus and an envelope protein, the surface antigen. Interestingly, the gene sequence for the surface protein is contained within the coding sequence for the polymerase, but in a different *reading frame*.

Some viruses produce large polyproteins, which are subsequently cleaved by proteases to produce the mature proteins of the virus. This is seen in HIV and so drugs that act as protease inhibitors, preventing the processing of the polyprotein, are used as part of the treatment given to HIV patients. Another interesting feature, illustrated by viruses like HIV, is the ability to copy RNA into DNA in a process called reverse transcription (see Chapter 5). The viral DNA then becomes inserted into the genomic DNA of the infected cell, where it can remain transcriptionally inactive and hence dormant for long periods of time, but it persists because the viral DNA is replicated along with the genomic DNA by the cell during cell division.

Viruses often have high spontaneous rates of mutation; this is particularly true for RNA viruses and is important in infection where mutations can allow the development of resistance to drugs or vaccines. The influenza virus is an RNA virus and is known to show gradual changes to the protein molecules on its surface (specifically haemagglutinin and neuraminidase) – these changes arise as a result of small point mutations occurring during replication. This process of gradual change via mutation is called antigenic drift and is the reason why influenza vaccines have to be updated each year.

CLINICAL EXAMPLE: FLU PANDEMICS

Significant changes can arise in the influenza virus via a process known as antigenic shift. The influenza virus has a genome that is divided between eight pieces of RNA, each encoding different characteristics of the virus. Antigenic shift occurs when two influenza viruses, typically a bird virus and a human virus, both infect an intermediate host (typically a pig) at the same time, allowing the genomes to mix and produce a new variant that contains some genetic material from each of the two viruses (Fig. 7.3). This new variant will not be recognised by the immune systems of either humans or birds and so can cause a pandemic. The Asian flu pandemic of 1957 is thought to have been caused by antigenic shift and killed about two million people worldwide.

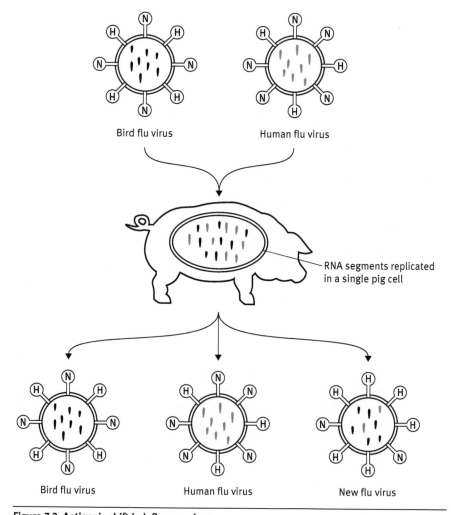

Figure 7.3. Antigenic shift in influenza virus

7.4 Fungi

Fungi are eukaryotic organisms, classified as a separate kingdom distinct from plants and animals. They comprise a diverse range of species including unicellular yeasts, moulds (such as those that appear on bread) and mushrooms/toadstools. Fungal cells possess membrane-bound nuclei and the typical cellular organelles of eukaryotes including mitochondria and ribosomes. The cells are enclosed by a cell wall made up mainly of chitin, a polysaccharide of *N*-acetylglucosamine. Most fungi grow as thread-like filamentous structures called hyphae, a network of hyphae making up the mycelium. Fungi can reproduce by either sexual or asexual means (see Section 11.1).

Fungi are a common cause of disease in plants, for example, potato blight, which was responsible for the Great Irish Famine in the 19th century. A few species can also cause disease in animals including humans. Some fungi infect the skin and cause conditions such as ringworm and athlete's foot, whereas others like *Candida* tend to infect mucosal membranes causing thrush. Fungi are frequently a cause of *opportunistic infections*, taking advantage of a weakened immune system. This is seen in patients with HIV who may succumb to pneumonia due to the fungus *Pneumocystis*, or to meningitis caused by *Cryptococcus*. Fungal spores can also be the cause of allergic reactions (see Section 25.4).

7.5 Other infectious agents

The largest infectious agents are the eukaryotic parasites. These can be single-celled (protozoa) or multicellular organisms including worms. Some of these parasites have complex life cycles involving multiple stages and different hosts. Malaria is a protozoal disease (caused by species of *Plasmodium*) and a major cause of death in some parts of the world. Sleeping sickness is a disease caused by another type of protozoa called a trypanosome. Both of these diseases are transmitted by insects: mosquitos in the case of malaria and tsetse flies in the case of sleeping sickness.

Schistosomiasis is a disease caused by a type of helminth worm of the genus *Schistosoma*. Schistosomes infect freshwater snails where they undergo replication and then are released into the water. Humans can acquire infection simply by coming into contact with the contaminated water because schistosomes are one of the few infectious agents that can penetrate intact skin.

7.6 Treatment of infectious disease

Treatment of infectious disease relies on drugs that selectively attack pathways that are necessary for the infectious agent but not for the healthy host. For example, many of the antibiotics that act against bacterial infection do so by inhibiting the synthesis of bacterial cell walls. Antiviral drugs may inhibit enzymes that are required for the synthesis of new virus such as reverse transcriptase. Eukaryotic infections can be more difficult to treat because many of the pathways used by the infectious agent are similar to those used by the host. As a consequence the drugs that are used may cause side effects. Certain antifungal drugs exert their action by binding to or inhibiting the synthesis of a molecule called ergosterol, which is found in fungal but not mammalian cell membranes.

CLINICAL EXAMPLE: MALARIA

Malaria is a major global disease which is estimated to kill at least 2000 people every day. It is caused by the protozoal parasite *Plasmodium* of which *Plasmodium falciparum* is the most common species. This parasite lives within the body of the female *Anopheles* mosquito which acts as its primary host. The *Plasmodium* reproduces sexually in the gut of the mosquito and the resultant offspring (sporozoites) travel to the mosquito's salivary glands from where they are injected into the bloodstream of the individual (called a secondary host) who is bitten by the mosquito. Once in the bloodstream of the secondary host the sporozoites travel to the liver where they enter liver cells and reproduce asexually, producing thousands of cells known as merozoites. These cells leave the liver and infect red blood cells, dividing further to cause the red blood cells to rupture. This triggers the clinical symptoms of malaria, which include shivering, fever and sweating. Some of the merozoites produce gametocytes which remain within the bloodstream of the secondary host until they are taken up by a mosquito feeding on their blood. Once they enter the gut of the mosquito the male and female gametocytes fuse and new sporozoites are produced, ready to infect another secondary host.

Because of the complex life cycle of the *Plasmodium* parasite, elimination of malaria may depend on a combination of strategies including the disruption of mosquito breeding, the use of mosquito nets to limit access of the mosquito to humans at night when the mosquito feeds, and treatment of affected individuals with antimalarial drugs. Unfortunately, many *Plasmodium* species are now developing resistance to the most commonly available drugs such as chloroquine and so new therapies are being developed.

7.7   Test yourself

The answers are given on p. 175.

Question 7.1
What is the major constituent of bacterial cell walls?

Question 7.2
What term is used to describe the process by which some RNA viruses are able to produce DNA copies of their genome?

Question 7.3
Some viruses can infect bacteria. True or false?

Question 7.4
Which polysaccharide is the major constituent of fungal cell walls?

Question 7.5
What type of organism causes malaria?

08 Energy metabolism

BASIC CONCEPTS:
Cellular metabolism is driven by the energy stored in the ATP molecule. Without energy, cells will stop functioning and it is therefore important to understand the factors which regulate its production. ATP is produced by using the energy derived from the breakdown of complex organic molecules. To break down these molecules fully requires the presence of oxygen, which is combined with the carbon and hydrogen to form carbon dioxide and water. This process is known as aerobic respiration and takes place within the mitochondria.

8.1 Introduction

We derive the energy necessary for life from the various foods we eat. Ultimately the energy stored in food is derived from the sun through the actions of plants which utilise special organelles known as chloroplasts to combine carbon dioxide and water into carbohydrates. In order for us to release the energy stored in foodstuffs they must be broken down. Energy released during these reactions is transferred to carrier molecules of ATP (*adenosine triphosphate*). In ATP the third phosphate group is bound by a high-energy bond (Fig. 8.1), and this energy can be released as required by the conversion of ATP into ADP (*adenosine diphosphate*).

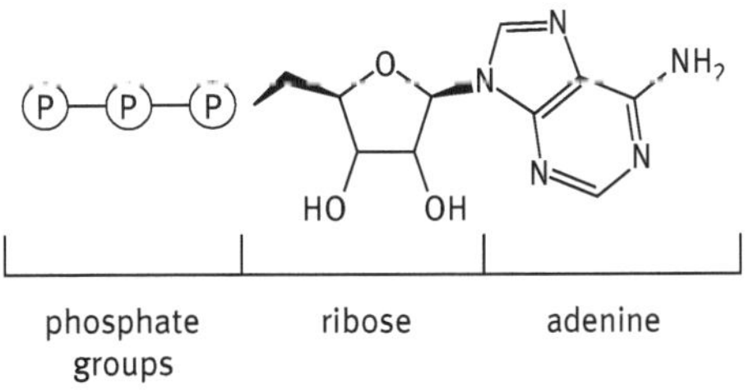

Figure 8.1. Structure of ATP

The breakdown of glucose gives rise to large quantities of ATP. Glucose derived from food is broken down in a step-by-step process into carbon dioxide and water. The reaction uses oxygen and can be represented as:

$$C_6H_{12}O_6 + 6O_2 \rightarrow 6CO_2 + 6H_2O.$$

This does not occur as a single chemical reaction but in an ordered series of reactions in which the energy from glucose is progressively released and stored in carrier molecules.

The first series of reactions occurs in the cytoplasm and is described as *glycolysis*. In glycolysis the six-carbon sugar glucose is broken down into two molecules of the three-carbon compound *pyruvate*. This occurs as a series of ten separate reactions, each requiring its own enzyme. Such a series of reactions, in which the product of each reaction forms the substrate for the next, is called a *metabolic pathway*. The principle is illustrated in Fig. 8.2. For every molecule of glucose that is converted to pyruvate there is an associated release of energy in the form of two molecules of ATP. Additionally glycolysis leads to the production of NADH from NAD^+ (nicotine adenine dinucleotide). NADH serves as a carrier of high-energy electrons and their energy is subsequently used to generate more ATP (see below).

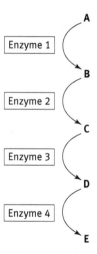

Figure 8.2. Representation of a metabolic pathway

8.2 The citric acid cycle

The citric acid cycle (also known as Krebs' cycle or the tricarboxylic acid (TCA) cycle) is a metabolic pathway whose enzymes are located in the mitochondrial matrix. It generates CO_2 and NADH.

Pyruvate produced as a result of glycolysis in the cytoplasm is transported to the mitochondria, where it is combined with coenzyme A (CoA) to form a molecule called acetyl CoA, with the release of carbon dioxide and the production of another molecule of NADH. The acetyl group is transferred from acetyl CoA to a four-carbon molecule called oxaloacetate to produce the six-carbon molecule citric acid and the CoA is released. Citric acid then forms the starting point of a series of reactions in which the two carbons derived from the acetyl group of acetyl CoA are oxidised into carbon dioxide. The end product of these reactions is oxaloacetate, which can combine with more molecules of acetyl CoA so that the cycle continues. As a result of the citric acid cycle more NADH is produced.

8.3 Release of energy from fats and proteins

So far the discussion has focused on the release of energy from carbohydrates, particularly glucose. Fats are also a major source of energy and these are digested in the gut to release fatty acids. Acetyl CoA can be produced in the cell from fatty acids and this then enters the citric acid cycle as described above. Similarly proteins are digested into amino acids and these can be further broken down to feed the citric acid cycle and generate energy.

CLINICAL EXAMPLE: BERIBERI

Deficiency of vitamin B1, also known as thiamine, leads to a condition called beriberi, which mainly affects the nervous system. Symptoms may include numbness, pain and paralysis. Thiamine is required for the action of the enzyme pyruvate dehydrogenase, which is involved in converting pyruvate produced from glucose in glycolysis into acetyl CoA, which can then enter the citric acid cycle. Why is the nervous system selectively affected by this? Unlike other cells of the body which can also use fats and proteins, it appears that nerve cells depend on glucose as their main source of fuel. So when pyruvate dehydrogenase is inactive this leads to neurological problems.

8.4 Oxidative phosphorylation

The high-energy electrons in the NADH, generated as a result of glycolysis and the citric acid cycle, are then used to generate more ATP. The electrons are transferred in an electron transport chain down a series of carriers. During this process their energy is released and used to pump hydrogen ions (H^+) out of the mitochondrial matrix into the space between the inner and outer mitochondrial membranes to create a concentration gradient. This gradient is then used by the ATP synthase enzyme to produce ATP. This enzyme is located in the inner mitochondrial membrane and as hydrogen ions pass back down their concentration gradient through a narrow channel in the enzyme the energy of their flow is used to drive the production of ATP. This can be viewed as the molecular equivalent of the way in which water flowing over a waterwheel is used to generate electricity. This process is termed oxidative phosphorylation (Fig. 8.3). Ultimately the electrons and H^+ ions are transferred to oxygen to produce water in an overall reaction that can be described by the equation

$$NADH + \tfrac{1}{2}O_2 + H^+ \rightarrow NAD^+ + H_2O$$

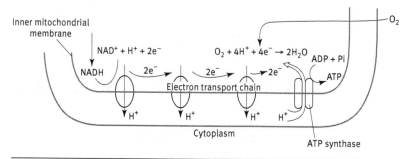

Figure 8.3. Oxidative phosphorylation in the mitochondrion

Overall each molecule of glucose generates about 30 molecules of ATP. Two molecules of ATP are produced directly from glycolysis, but the majority of the ATP is produced in the process of oxidative phosphorylation using the NADH generated in glycolysis and the citric acid cycle (Fig. 8.4).

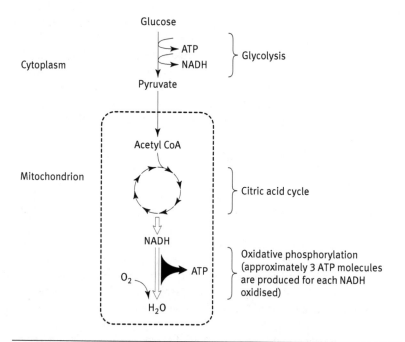

Figure 8.4. Relationship between glycolysis, the citric acid cycle and oxidative phosphorylation

8.5 Anaerobic respiration

The complete breakdown of glucose as described above requires oxygen.
Under *anaerobic* conditions, where oxygen is deficient, *glycolysis* is the main
source of ATP. Anaerobic conditions can arise in muscle tissues as a result of
oxygen depletion in exercise. In this case the pyruvate produced in glycolysis
is converted into lactic acid. This process uses electrons donated by NADH,
which is converted back to NAD^+.

8.6 Test yourself

The answers are given on pp. 175–176.

Question 8.1
What three-carbon compound is produced from
glucose as a result of glycolysis?

Question 8.2
What does the abbreviation ATP stand for?

Question 8.3
Where in the cell does the citric acid cycle
occur?

Question 8.4
What term is used to describe the process
whereby a proton gradient across the inner
mitochondrial membrane is used to drive the
synthesis of ATP?

Question 8.5
In the absence of oxygen into what is pyruvate
converted?

09 Membrane transport

> **BASIC CONCEPTS:**
> The cell membrane is semi-permeable, which means that it will permit some substances to pass through it more easily than others. Cell functions depend on the passage of water, ions and metabolites into and out of the cell. This can be achieved through simple diffusion or with the help of specific proteins in the membrane. Understanding the mechanisms of membrane transport is essential for understanding many basic functions of the body.

9.1 Introduction

All cells need to exchange material with the extracellular fluid which bathes them. This may include nutrients, oxygen, carbon dioxide, waste products, hormones, ions or other cellular secretions. The plasma membrane is selectively permeable to these various substances and the mechanisms by which they enter or leave the cell will depend on their permeability.

Small molecules such as O_2, CO_2 and urea are freely diffusible across the plasma membrane. Their transport follows the laws of diffusion in that the rate of movement depends primarily on the concentration difference between the inside and outside of the cell. Gaseous exchange in the alveoli of the lung (see Chapter 21) is an example of this mechanism. Lipid-soluble molecules such as steroid hormones will also pass freely through the plasma membrane.

Other molecules are not able to diffuse across the cell membrane and are transported by specialised mechanisms involving membrane-associated transport proteins. Where molecules are moving down a concentration gradient this process is described as facilitated diffusion or passive transport. In order to move molecules against a concentration gradient – from areas of low concentration to areas of high concentration – energy is required and a process of active transport is involved.

9.2 Osmosis

Perhaps surprisingly, given the hydrophobic nature of the centre of the lipid bilayer, water can easily pass through the plasma membrane. This is due to the small size and lack of overall charge of the water molecule. The movement of water across plasma membranes is achieved by a process known as *osmosis* (Fig. 9.1) which occurs when two basic conditions are satisfied:

- there must be a higher concentration of particles of solute on one side of the membrane than the other;
- the membrane must be more permeable to water than to the solute particles.

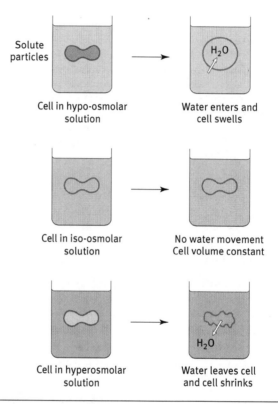

Figure 9.1. Osmosis and its effects on cells

If these conditions are met then water will move from an area of low solute concentration to one of high solute concentration.

The *osmolarity* of a solution is a measure of the number of particles dissolved in it and is determined by the molarity of the solute (see Section 4.3 in *Catch Up Chemistry*) multiplied by the number of particles into which it dissociates. Osmolarity is expressed in the units osmol/l. Thus a 1 molar solution of sucrose has an osmolarity of 1 osmol/l as each sucrose molecule in solution is a single particle. However, a 1 molar solution of NaCl has an osmolarity of 2 osmol/l because in solution each molecule of NaCl dissociates into the two osmotically active particles Na^+ and Cl^-. Both intracellular and extracellular fluids in the body have an osmolarity of about 290 mosmol/l and an important function of homeostasis (see Chapter 17) is to maintain this value. Solutions that have the same osmolarity are said to be *iso-osmolar*.

An alternative way of expressing osmolarity is to use the term *osmolality*. This is a measure of the number of particles of solute per mass of solvent rather than its volume (i.e. the units are mosm/kg H_2O). At standard temperature and pressure for an aqueous solution containing only one solute at low concentration, the values for osmolarity and osmolality will be identical. This is because the volume occupied by the solute is negligible and a litre of solution will contain 1 kg of H_2O. However, in the measurement of osmolality in biological fluids such as plasma, it is necessary to take into account the volume occupied by other solutes such as protein. Thus 1 litre of plasma will weigh 1 kg but will actually only contain 930 g of H_2O (total plasma protein = approx 70 g). When calculating osmolality of Na^+ in a litre of plasma we calculate not as if this litre was equal to 1 kg of H_2O but instead as if it were 930 g H_2O. This means that a solute with a calculated osmolarity in plasma of 100 mosmol/l will have a higher calculated osmolality since it is actually distributed in less water. **The terms osmolarity and osmolality are often used interchangeably and in most circumstances this lack of precision has little functional significance.**

Solutions that have the same osmolality are said to be iso-osmolal. A solution with a lower osmolality than another is hypo-osmolal and one with a higher osmolality is hyperosmolal. A difference in osmolality between two solutions which leads to the movement of water is described as the effective osmolality or tonicity. The movement of water into a compartment generates osmotic pressure. A difference in effective osmolality of 1 mosmol/kg is equivalent to a pressure of 19.3 mmHg. If cells which are surrounded only by a plasma membrane are placed into pure water osmosis will cause the movement of water into the cells, raise their internal pressure, and they will swell until they burst, a process known as lysis (Fig. 9.1).

The internal pressure generated by the movement of water into plant cells (which have a rigid cellulose cell wall and therefore cannot swell) is known as turgor and helps to prevent the leaves and stems of plants from going floppy. (Think what happens to cut flowers if they are left out of water!)

The movement of water between the various compartments of the body is controlled predominantly by the movement of ions. For example, the secretion of water from the cells of the epithelium lining the airways in order to reduce the stickiness of the mucus which coats them is achieved by moving sodium ions and chloride ions out of the airway cells. This increases the extracellular solute concentration and water follows by osmosis.

The movement of water back into capillaries from the interstitial fluid is also due to osmosis, but in this case the key particles are the proteins within the capillary. In many tissues the capillary walls are as freely permeable to ions as they are to water and so the ions exert no osmotic influence – the first condition required for osmosis to occur is not satisfied. However, the proteins are too large to move from the capillary and so are osmotically active and because the interstitial fluid contains no protein there is a concentration gradient of solute which draws water into the capillary (see *Oncotic pressure* in Chapter 20).

9.3 Facilitated diffusion

Many substances need to be transported into and out of cells but have low permeability in the cell membrane and so will not readily diffuse across. This is true of all ions and of larger organic molecules such as glucose. Where such substances have a concentration gradient across the membrane then their transport can occur through a process known as *facilitated diffusion* – also known as *passive transport*. This can occur with the assistance of either *carrier molecules* or *channels*.

Carrier molecules are trans-membrane proteins which bind the molecule to be transported on the side of the membrane where it is at high concentration and then undergo a conformational change which moves the molecule through the membrane before releasing it on the other side. The carrier protein then returns to its original shape and is ready to transport another molecule. Some carriers only move one molecule or ion through the membrane and are known as uniports. Others transport two different molecules or ions in opposite directions across the membrane and are known as antiports.

Channels are small protein-lined openings in the membrane which permit the movement through the membrane of a molecule down its concentration gradient. Channels have two key properties in that they are regulated and selective. Channels may exist in either open or closed states and the probability of being in one of those states depends on external factors. Thus some channels will tend to open when a signal molecule, such as a neurotransmitter, binds to the proteins which make up the channel. Other channels are voltage sensitive and will open when the membrane potential (see Chapter 16) moves above a certain threshold. The size of the channel opening and the charge on the proteins which make up the channel lining will determine which molecules may pass through. Many channels are only large enough to permit the passage of ions and will be selective for particular cations or anions such as Na^+ and Cl^-.

9.4 Active transport

In both diffusion and facilitated diffusion, substances move across the membrane down their concentration gradient. It is also possible to use energy directly or indirectly to move substances up their concentration gradient.

In primary *active transport* the energy derived from the breakdown of ATP to ADP is used directly to pump substances across a membrane against their concentration gradients. One of the most widespread of these pumps is the Na^+/K^+ ATPase. This protein complex pumps three Na^+ ions out of the cell and two K^+ ions into the cell. This helps to maintain the high extracellular Na^+ and high intracellular K^+ levels which are characteristic of eukaryotic cells. This process is so important to cell function that approximately 30% of all ATP generated is used to fuel this pump. Other pumps important to cellular function are those that pump Ca^{2+} ions out of the cytoplasm of excitable tissues once the depolarising event which has triggered Ca^{2+} entry into the cytoplasm is complete.

CLINICAL EXAMPLE: CHOLERA

Infection by the bacterium *Vibrio cholerae* causes the disease cholera which is characterised by watery diarrhoea and potentially fatal dehydration. The bacteria release a protein toxin which enters cells in the intestine and causes the opening of channels in their membranes. This allows Cl^- ions to flow from the cells into the lumen of the gut. This movement of ions draws water with it by osmosis, leading to diarrhoea and producing dehydration in the rest of the body. This is potentially fatal. The most effective way of rehydrating the body is to drink a mixture of glucose and salts dissolved in water. This mixture activates co-transporters in the gut wall, which take up glucose in the presence of Na^+ ions. The inward movement of Na^+ ions and glucose moves water by osmosis back from the lumen of the gut into the bloodstream and reverses the effects of the toxin.

Secondary active transport depends on the maintenance by the Na^+/K^+ ATPase of an Na^+ ion gradient between the outside of the cell and the inside. The movement of Na^+ ions into the cell down their concentration gradient is then coupled with the transport by a carrier protein (co-transporter or symport) of a substance into the cell against its concentration gradient. This mechanism is used in the transport of sugars, amino acids, nucleotides and other essential nutrients into cells. The Na^+ ions that enter the cell are immediately recycled out of the cell by the Na^+/K^+ ATPase and the Na^+ ion gradient is thus maintained to drive further transport activity.

9.5 Exocytosis and endocytosis

Transfer of very large molecules, such as proteins, across the cell membrane is achieved not by the transport of individual molecules using carrier proteins but by the mechanisms of *exocytosis* and *endocytosis*. In the process of exocytosis, proteins which are destined for export from the cell are packaged in the Golgi apparatus into small vesicles. These vesicles then move towards the area of the cell membrane from which they are to be secreted and are held next to the membrane by linker proteins. When the appropriate signal is received a Ca^{2+}-mediated mechanism triggers the fusion of the vesicle with the cell membrane and the contents of the vesicle are emptied into the interstitial fluid. Exocytosis is used for the storage and release of hormones into the bloodstream (see Chapter 18) and digestive enzymes into the lumen of the gut (see Chapter 22).

Endocytosis is the reverse of exocytosis and involves the binding of substances to be taken into the cell by receptors on the cell surface. Once the receptors are occupied then the membrane is drawn in and a vesicle is pinched off. Such vesicles are most frequently directed to the lysosome where the ingested material is broken down.

9.6 Test yourself

The answers are given on p. 176.

Question 9.1
What is the osmolarity of a 1 molar solution of $CaCl_2$?

Question 9.2
True or false: passive transport can only occur down a concentration gradient?

Question 9.3
The movement of amino acids into a cell against their concentration gradient is coupled with the movement of which cation into the cell?

Question 9.4
What is the name of the enzyme responsible for maintaining high extracellular concentrations of Na^+ ions and high intracellular concentrations of K^+ ions?

Question 9.5
Is urea osmotically active in the body?

10 Cell division and mitosis

> **BASIC CONCEPTS:**
> The growth of organisms and the replacement of cells both depend on the formation of new cells. These arise by a process of division, either from differentiated cells within a tissue or from stem cells. In order to divide, cells have to make a copy of each of their chromosomes so that the daughter cells can be genetically identical to the parent cell. The sequence of events by which a cell reproduces itself is known as the cell cycle. Understanding the cell cycle and its regulation allows us to understand what may go wrong when uncontrolled cell division produces tumours.

10.1 Introduction

For a multicellular organism to grow and for tissues to be regenerated requires an increase in the number of cells. This takes place through an orderly process of events involving the duplication of the genetic material and separation of this into two new daughter cells.

10.2 Cell cycle

The stages through which a cell proceeds as it divides are described as the *cell cycle* (Fig. 10.1). Replication of the cell's genetic material, the DNA, is described in Chapter 5 and this occurs during the S-phase of the cell cycle. At the end of the S-phase a cell therefore has twice its normal amount of DNA, with two copies of each chromosome. Each copy is known as a *chromatid* and the two chromatids are joined by a structure known as the *centromere*. During the M-phase of the cell cycle this genetic material is carefully segregated into two new daughter cells, each of which inherits one copy of each chromosome; a process known as mitosis (Fig. 10.2).

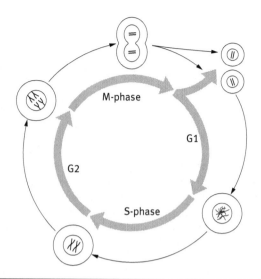

Figure 10.1. The cell cycle

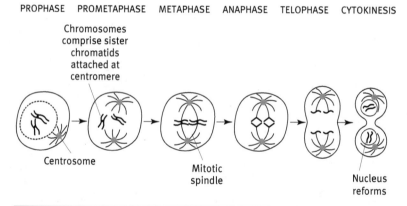

Figure 10.2. The phases of mitosis

After the M-phase and before the cell enters the S-phase and starts to replicate its DNA again there is a period described as G1. Similarly after the S-phase and before the M-phase there is a second period called G2. G1 and G2 provide checkpoints where the cell can decide whether or not to continue on to the next phase of the cell cycle. Together G1, S and G2 are described as *interphase*. G1 is also important as the period during which cell growth occurs. One of the factors which would prevent the cell entering the S-phase and proceeding through a further mitotic division is small size. Thus, cells spend longer in G1 in malnourished individuals and if this occurs during development it can permanently reduce the eventual cell population of organs such as the brain.

The M-phase can be subdivided into six phases, each of which is characterised by particular events.

- In *prophase* the chromosomes condense and the *mitotic spindle* starts to form in the cytoplasm. The mitotic spindle is made up of various microtubules and, in animal cells, involves an organelle called the *centrosome*, which duplicates and gives rise to the two poles of the spindle.
- In *prometaphase* the nuclear membrane starts to break down so that the chromosomes can now become attached to the mitotic spindle. The sister chromatids of the duplicated chromosomes become attached to the spindle via the centromere.
- In *metaphase* the chromosomes arrange themselves in the centre of the cell equidistant from each pole of the spindle.
- In *anaphase* the sister chromatids now separate, forming two daughter chromosomes, and these are pulled towards the poles of the spindle (one of each pair going to each pole).
- There are now two identical sets of chromosomes within the cell, one at each end. In *telophase* a nuclear envelope is formed around each of the two sets of chromosomes
- Finally the cytoplasm divides to give rise to two daughter cells: this is known as *cytokinesis*. Chromosomes within the two daughter cells now de-condense and cell division is complete.

10.3 Control of cell division

Within the body, cells in different tissues tend to divide at very different rates. Cells in the gut are dividing rapidly to replace cells that are constantly being lost. In contrast, nerve cells are essentially nondividing cells. Cell division is strictly regulated by a series of molecular controls and if these go wrong then uncontrolled cell division can occur, leading to the development of cancers.

The cell cycle is safeguarded by *checkpoints*. The first checkpoint comes at the end of G1 – at this point cells are allowed to enter the S-phase (DNA synthesis) only if conditions in the environment of the cell are appropriate and the DNA is not damaged. The second checkpoint occurs at the end of G2 and only allows the cell to enter the M-phase if the DNA has replicated correctly. The third checkpoint occurs during the M-phase and checks that all of the chromosomes have been correctly arranged on the spindle during metaphase before they separate during anaphase. Together these checkpoints ensure that cell division only occurs under appropriate conditions and that each daughter cell inherits a complete and undamaged copy of the genetic material. These controls depend on enzymes called *cyclin-dependent kinases*, which phosphorylate proteins that regulate the cell cycle.

Cell division is also regulated by factors in the external environment. In mammals there is a group of proteins called *growth factors* that bind to cell surface receptors and stimulate cells to divide. Growth factors can either be broadly specific for a range of tissues or act on a very narrow range of cell types. Erythropoietin is an example of a growth factor that is highly specific and stimulates the production of red blood cells (see Chapter 20). This property has been exploited by some athletes who have used erythropoietin as a performance-enhancing drug.

Many human cells are limited in the number of rounds of cell division they can undertake. As they divide, material is progressively lost from the ends of the chromosomes (called *telomeres*). This shortening of the telomeres ultimately prevents further cell division occurring. This is thought to provide protection against unregulated cell growth and the development of cancer.

CLINICAL EXAMPLE: CANCER AND P53

Many human cancers are associated with mutations in a gene called *p53*. The p53 protein plays an important role in the cell cycle at the checkpoint between G1 and entry into the S-phase. If DNA damage is detected this leads to the expression of p53 which then blocks DNA replication until the damage has been repaired. Mutations leading to a loss of p53 function will allow cells with damaged DNA to divide and to accumulate further mutations. These mutations, by disrupting the functions of other important genes, can lead to the unregulated cell division that is characteristic of cancer. The p53 gene is therefore described as a *tumour suppressor gene*.

10.4 Cell division and differentiation

As cells divide they give rise to more cells of the same type. For example, a dividing epithelial cell will give rise to further epithelial cells. These *differentiated* cells maintain their properties due to the regulation of gene expression (see Chapter 5) which ensures that a set of proteins is produced that is characteristic of that cell type.

When they reach their fully differentiated state some cells lose the ability to undergo further cell division and are said to be *terminally differentiated*. An example of this is red blood cells which have no nucleus and can no longer divide. As red blood cells die they are replaced from *stem cells*, which give rise to the various cells of the blood. Stem cells are not differentiated but when they divide they give rise both to differentiated cells (e.g. red blood cells) and more stem cells. The stem cells in various tissues in an adult have a limited potential to give rise to different cell types. For example the haematopoietic stem cells that occur in the bone marrow give rise to red blood cells, white blood cells and platelets.

Researchers are very interested in the potential for using stem cells to replace damaged tissues in certain diseases. Of particular interest are the stem cells that can be found in early embryos (*embryonic stem cells*). These have the potential to give rise to any cell type. In the future, it is hoped that embryonic stem cells might be used to replace damaged tissues and restore normal function in patients with a range of diseases.

10.5 Test yourself

The answers are given on p. 176.

Question 10.1
What three phases of the cell cycle comprise interphase?

Question 10.2
What is the name given to the structure formed from microtubules that moves duplicated chromosomes to opposite poles of the cells?

Question 10.3
What is the region called in a mitotic chromosome where the two sister chromatids are attached to each other and to the spindle?

Question 10.4
What name is given to the regions at the ends of chromosomes which tend to shorten as cells undergo successive rounds of cell division?

Question 10.5
What are the enzymes called that regulate progression through the cell cycle by phosphorylating other proteins?

⬡11 Reproduction

> **BASIC CONCEPTS:**
> The purpose of reproduction is to ensure the survival of the species by the production of new individuals. Asexual reproduction results in offspring that are identical to the parent. Sexual reproduction leads to the mixing of genes from two parents to form a unique organism. Because evolution depends on variability within a population it is important that during the production of the egg and sperm there is a random allocation of genetic material from each of the parents. This randomisation, which occurs during meiosis, ensures that every egg and sperm is genetically unique and that no two individuals born of the same parents will be identical. Understanding this process allows us to appreciate how individuality is created at the genetic level and how a species becomes equipped to face multiple, unknown challenges.

11.1 Introduction

An essential characteristic of living organisms is the ability to reproduce. Reproduction can be achieved in a number of different ways, some of which are more complex than others. Single-celled organisms such as bacteria tend to reproduce by cell division. Such reproduction is described as *asexual* (not involving sex). In asexual reproduction only a single parent is required and the progeny are identical to the parent. Asexual reproduction is also seen in more complex, multicellular organisms. Plants have a variety of means of asexual reproduction involving, for example, the production of tubers in potatoes, bulbs in flowering plants such as daffodils, runners in strawberries etc. Some animals can also reproduce asexually, for example the jellyfish.

Most plants and animals use sexual means of reproducing themselves, either instead of or in addition to asexual reproduction. *Sexual reproduction* involves two parents and the offspring inherit genes from both (Fig. 11.1). The genes are inherited in various combinations so that the offspring will be different from each other and from the two parents. This increased variability is thought to be advantageous in allowing a species to adapt to the environmental conditions.

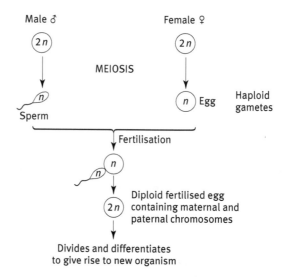

Figure 11.1. Sexual reproduction results in offspring with genetic material derived from both parents

11.2 Sexual reproduction

Gametogenesis

Organisms that reproduce through sexual means are *diploid*, that is to say each cell contains two copies of each chromosome – one inherited from the mother (*maternal*) and the other from the father (*paternal*). In humans there are 23 pairs of chromosomes. The two chromosomes of a pair are said to be *homologous* and each carries the same set of genes, which means that each gene is present in two copies.

In order to reproduce sexually the organism must produce *germ cells* or *gametes*. Gametes are *haploid*, which is to say they carry only a single set of chromosomes. In order to produce haploid gametes, cells in the reproductive organs must undergo a special type of cell division called *meiosis*. Because meiosis leads to the production of cells which have only half as many chromosomes as the cell from which they are derived this is sometimes called *reductive division*. The germ cells of the male are described as *spermatozoa*, or *sperm*, and the germ cells of the female are called *ova*, or *eggs*. The process through which eggs and sperm are produced is called *gametogenesis*. In the process of fertilisation, male and female gametes fuse to form a zygote, which is now diploid, having acquired one set of chromosomes from each gamete.

Meiosis

The process of meiosis shares some similarities with mitosis but also shows some important differences. DNA is duplicated in mitosis and meiosis but in the latter there are two, rather than just one, rounds of cell division with the result that each cell inherits half the DNA of the parent cell.

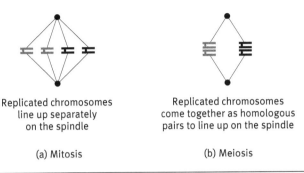

Replicated chromosomes line up separately on the spindle	Replicated chromosomes come together as homologous pairs to line up on the spindle
(a) Mitosis	(b) Meiosis

Figure 11.2. Pairing of homologous chromosomes during metaphase in meiosis compared with metaphase in mitosis

At the start of meiosis, as in mitosis, the chromosomes condense, the nuclear membrane breaks down and the spindle appears. DNA undergoes replication resulting in the production of two sister chromatids that stay closely associated with each other. This stage is described as *prophase I*. The next event is unique to meiosis and is not observed in mitosis. Each replicated chromosome comes together with its homologous pair (Fig. 11.2). This pairing of homologous chromosomes is essential for the next stage because it allows each daughter cell to inherit one member of the pair.

Pairing of homologous chromosomes also allows for recombination to occur. In the process of recombination, or *crossing-over* (Fig. 11.3), there is a

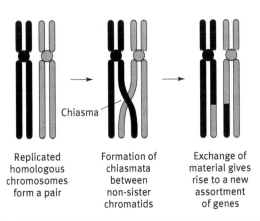

Replicated homologous chromosomes form a pair	Formation of chiasmata between non-sister chromatids	Exchange of material gives rise to a new assortment of genes

Chiasma

Figure 11.3. Crossing-over between homologous chromosomes at meiosis

reciprocal exchange of material between homologous chromosomes. Thus the paternal chromosome in a pair donates some material to the maternal chromosome and acquires in exchange some of the maternal chromosome. Crossing-over requires the formation of a connection between two non-sister chromatids and this is known as a *chiasma* (plural *chiasmata*). On average each pair of human chromosomes will form 2–3 chiasmata. This phase is described as *anaphase I*.

The next step is for the duplicated homologous chromosomes (*homologues*), each consisting of two chromatids, to separate. These are pulled to opposite poles of the spindle and then cell division occurs so that each daughter cell inherits a set of duplicated chromosomes. The daughter cells now enter into a second round of cell division. Spindle formation occurs in each daughter cell (*prophase II*) and then the chromosomes line up on the equator (*metaphase II*). At this point, the sister chromatids now separate and are pulled to opposite poles. This stage is described as *anaphase II* and is followed by reformation of the nuclear membrane and cell division.

CLINICAL EXAMPLE: DOWN SYNDROME

Errors sometimes occur during meiosis such that the chromosomes separate incorrectly – this is known as *non-disjunction* – and can result in the production of gametes that lack a particular chromosome or have two copies. Such an error occurs in Down syndrome, which results from an extra chromosome 21 (trisomy 21). Down syndrome arises when one of the gametes, usually the egg, acquires two copies of chromosome 21 so that after fertilisation the resulting embryo has three copies of this chromosome. This results in abnormal development and affected children show learning disabilities and characteristic facial features. The condition is also associated with an increased risk of infection and, in some cases, heart problems.

Down syndrome occurs more commonly in the children of older mothers and can be detected by examination of fetal cells following amniocentesis.

In this way a diploid cell undergoes one round of DNA replication and two rounds of cell division resulting in four gametes, each with half the DNA of the original cell. Each gamete has a full set of chromosomes (23 in humans) with a mixture of maternal and paternal homologues. The random shuffling of maternal and paternal homologues together with the crossing-over that occurs in the first cell division leads to enormous genetic variability between gametes.

11.3 Fertilisation

Fertilisation involves the fusion of two haploid gametes to produce a diploid zygote. Spermatozoa, the male gametes, are produced in the testes (singular *testis*; see Chapter 23) in the process of *spermatogenesis*. A sperm is a very small specialised cell comprising a head, where the nucleus is located, and a tail to enable motility. There is also a special organelle at the tip of the head called the *acrosome*, which is derived from the Golgi apparatus and is involved in fertilisation. Mitochondria in the tail provide the energy to allow the sperm to swim in search of the egg.

Eggs (*oocytes*) are produced in the *ovaries* (see Chapter 23) in the process of *oogenesis*. These are large spherical cells and when they are shed from the ovaries they are surrounded by a special coating called the *zona pellucida* and a layer of cells from the ovarian follicle in which they developed.

For fertilisation to occur a sperm must contact the egg and then penetrate the outer layer of cells and the zona pellucida before fusing with the plasma membrane of the egg. Sperm are produced in large numbers and several can bind to an egg but it is crucial that only one sperm is allowed to fertilise the egg. To ensure this, once fertilisation has occurred an increase in intracellular Ca^{2+} ions leads to changes in the zona pellucida that prevent further sperm from penetrating. Within the fertilised egg the two nuclei must then fuse to form a diploid nucleus and the process of *embryogenesis* leading to the development of a new individual can begin. The cells divide and then implant themselves into the wall of the *uterus*, where the developing embryo can obtain nutrients from the mother's blood via the *placenta* (see Chapter 23). In humans the period from fertilisation to birth is 40 weeks (the *gestation period*).

11.4 Reproductive and therapeutic cloning

The term *clone* refers to a group of cells or organisms that originate from and are identical to a single precursor cell or organism. Asexual reproduction leads to the production of clones with an identical genetic make-up to the parent. Cloning has also been achieved in mammals by artificial means, and the best-known example of such *reproductive cloning* is Dolly the sheep. Reproductive cloning is achieved by taking an unfertilised egg, removing the nucleus and replacing it with the nucleus of a diploid cell from another animal. In the case of Dolly the nucleus was taken from an epithelial cell in the mammary gland. The egg is then allowed to develop in culture where it may give rise to an early embryo, which can then be implanted in the uterus of a foster mother, where it develops into a cloned animal. The animal will be genetically identical to the animal from which the nucleus was taken.

Therapeutic cloning is still under development. It uses a similar approach to give rise to embryonic stem cells of a defined genetic make-up. In this case, instead of the early embryo being implanted in a foster mother, it is used instead as a source of embryonic stem cells (see Chapter 10). Because these cells have the potential to differentiate into different tissue types there is hope that they could be used to help replace damaged tissues in a wide range of different diseases.

11.5 Test yourself

The answers are given on p. 176.

Question 11.1
What is the general term used to describe the haploid germ cells produced in sexual reproduction?

Question 11.2
What type of cell division results in the production of haploid cells?

Question 11.3
What name is given to the points on the chromatids at which crossing-over occurs?

Question 11.4
What is the layer surrounding the egg that must be penetrated by the sperm in order for fertilisation to occur?

Question 11.5
Therapeutic cloning aims to produce what type of cells in culture?

12 Inheritance

> **BASIC CONCEPTS:**
> Genes provide the information that determines the overall appearance and
> characteristics of an organism. The laws of inheritance allow us to predict
> how the genes inherited from a father and a mother will interact to produce
> the complex pattern of features that make up an individual. Some inherited
> traits, such as eye colour, are explained by the action of a few genes; others,
> such as personality, are the product of many genes and also of non-genetic,
> environmental factors. Understanding how inheritance works is essential in
> understanding the basis of genetic disease.

12.1 Introduction

It is widely recognised that offspring have a tendency to resemble their
parents. This is because the genetic material, DNA, is passed from one
generation to the next. A diploid organism such as a human being inherits
one set of chromosomes from each parent and the genes that are inherited
as part of the chromosomes account for many characteristics of the offspring.

Variation within a population can be due to either genes or environment, or a
combination of both. For example, tall parents will pass on the genes for
tallness to their children – however, if the children are poorly nourished this
can restrict their height. Some characteristics that are inherited are due to
single genes while other characteristics represent the combined effects of
multiple genes (these traits are said to be *polygenic*).

12.2 Definition of terms

The two copies of a gene that are inherited from the two parents are called
alleles. For any gene, if the two alleles that are inherited are identical then
that individual is said to be *homozygous*. If the two alleles that are inherited
are different then the individual is *heterozygous*. The position occupied by a
gene on a chromosome is described as a genetic *locus*.

The set of genes that is inherited by an individual is described as the
genotype. The expressed characteristics (for example height, hair colour,
blood group) constitute the phenotype. In a heterozygous individual one
allele can be *dominant* while the other allele is *recessive* – in this case the

phenotype represents the dominant allele. An example of dominance is seen in rhesus blood groups, where individuals who have either one or two copies of the gene for rhesus protein are rhesus positive, and only individuals who are homozygous for the recessive allele are rhesus negative. Some genes can be expressed co-dominantly, for example human cells have molecules on their membranes called *MHC* molecules. There are a number of different MHC alleles in human populations (such genes are said to be highly *polymorphic*) and so individuals are frequently heterozygous at these loci. A heterozygous individual will express on their cells the MHC proteins corresponding to the genes inherited from both the father and the mother.

12.3 Mutations

The primary source of genetic variation is *mutation*. A mutation manifests as a change in DNA. Many mutations are *silent*, because they occur in non-coding regions or because the change does not affect the coding sequence. Other mutations will cause changes to the protein encoded by the mutated genes. It is only mutations that occur in the sex cells of an organism that will be inherited. Mutations in somatic cells will usually pass unnoticed, although sometimes they can have serious effects, e.g. giving rise to cancers.

Gene mutations can affect a single base pair (*point mutations*) or multiple base pairs. There are three main types of gene mutation:

*   *substitution* of one nucleotide for another
*   *deletion* mutations that remove nucleotides
*   *insertion* mutations where nucleotides are added.

Chromosomal mutations are also seen and Down syndrome, which is due to trisomy of chromosome 21, where the sufferers have three copies of chromosome 21 instead of the normal two, is an example (see Chapter 11).

Mutations occur spontaneously at a low rate but can be accelerated by agents such as radiation and exposure to certain chemicals.

12.4 Mendelian inheritance

Gregor Mendel was an Austrian monk whose work on plant breeding in the nineteenth century forms the basis of the science of genetics. Mendel's breeding experiments allowed him to deduce and formulate a number of genetic 'rules'. Mendel observed that inheritance was due to discrete particles – what we now call genes – and that these are passed unchanged from one generation to the next. Different genes encode for different characteristics and each organism has a pair of genes – in modern terminology, alleles – encoding a particular characteristic. The alleles of these genes *segregate* during gametogenesis so that each gamete has only one copy of a particular gene. If the genes reside on different (nonhomologous) chromosomes then their alleles show *independent assortment* – i.e. the inheritance of particular alleles at one genetic locus does not affect the inheritance of alleles at another locus.

12.5 Monohybrid inheritance

One of the characteristics studied by Mendel was the inheritance of height in pea plants. In pea plants height may show discontinuous variation, with peas being either tall or short (dwarf). The gene that determines this characteristic has two alleles – T, the allele for tallness, and t, the allele for dwarfness. Tall peas that breed true (i.e. always produce tall peas) when crossed with each other are homozygous for T (TT), while dwarf peas that breed true are homozygous for t (tt).

When these true-breeding tall peas and dwarf peas are crossed with each other, all of the offspring are tall. This is explained in Fig. 12.1.

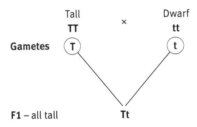

Figure 12.1. Outcome of crossing plants homozygous for the tall (T) and dwarf (t) alleles

All the offspring are heterozygous. The fact that they are all tall tells us that T is dominant to t. The offspring of this cross are described as the *F1* (standing for first filial) generation.

If a further cross is now performed between the F1 offspring, the pattern shown in Fig. 12.2 emerges.

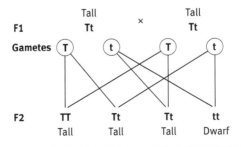

Figure 12.2. Outcome of crossing F1 generation plants heterozygous for the tall (T) and dwarf (t) alleles

Each F1 plant produces equal numbers of gametes containing either the T or t alleles. Random combination of gametes produces the next (F2) generation where the offspring comprise approximately 75% tall plants and 25% dwarf plants. The three possible genotypes, TT, Tt and tt will be represented in a ratio of roughly 1:2:1.

12.6 Dihybrid inheritance

Mendel also studied the simultaneous inheritance of two genes encoding different characteristics. In this case the genes were for height (tall or dwarf) and for flower colour (purple or white). When true-breeding tall purple-flowered plants were crossed with true-breeding dwarf white-flowered plants, all of the progeny were tall with purple flowers. We already know that the tall allele is dominant to dwarf – this experiment further shows that the allele for purple flowers is dominant to that for white flowers. The cross is illustrated in Fig. 12.3.

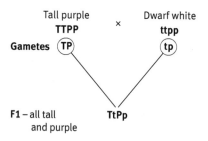

Figure 12.3. Outcome of cross between tall purple and dwarf white plants to illustrate dihybrid inheritance

The F1 plants are therefore heterozygous at both loci, and because the two genes are located on different chromosomes and their alleles assort independently, gametes produced by F1 plants can be of four types – TP, Tp, tP or tp – produced in equal numbers. When the F1 plants are crossed, a random combination of the four gametes from each parent produces the pattern shown in Fig. 12.4.

In other words, the following phenotypes are observed:

- 9 tall purple-flowered
- 3 tall white-flowered
- 3 dwarf purple-flowered
- 1 dwarf white-flowered.

F1		TtPp				TtPp		
Gametes		(TP) (Tp) (tP) (tp)				(TP) (Tp) (tP) (tp)		

F2		TP	Tp	tP	tp
		TTPP	TTPp	TtPP	TtPp
	TP	Tall, purple	Tall, purple	Tall, purple	Tall, purple
		TTPp	TTpp	TtPp	Ttpp
	Tp	Tall, purple	Tall, white	Tall, purple	Tall, white
		TtPP	TtPp	ttPP	ttPp
	tP	Tall, purple	Tall, purple	Dwarf, purple	Dwarf, purple
		TtPp	Ttpp	ttPp	ttpp
	tp	Tall, purple	Tall, white	Dwarf, purple	Dwarf, white

Figure 12.4. Outcome of cross between F1 generation plants heterozygous for tall, dwarf, purple and white

This shows that the two characteristics, height and flower colour, behave independently of each other, consistent with Mendel's law of independent assortment. It is worth noting that in the above cross the only plants whose genotype can be deduced from the phenotype are the dwarf white plants. Plants that are either tall or purple could be either homozygous or heterozygous for these genes because the alleles for these characters are dominant.

12.7 Linkage

Some pairs of genes are observed to disobey the law of independent assortment and tend to be inherited together. This is because they are located on the same chromosome, and such genes are said to be *linked*. Because crossing-over between homologous chromosomes occurs at meiosis (see Chapter 11) genes that are far enough apart on the same chromosome may also assort independently because of the high likelihood of recombination events occurring in the intervening DNA. Genes that lie close together are much less likely to behave independently as the probability of recombination is much lower. In this way, by observing how often recombination occurs between a pair of genes, we can estimate their relative distance from each other on the chromosome to give a *linkage map*. Such mapping has been important in helping to identify genes associated with several human diseases.

12.8 Autosomal and sex-linked genes

Human beings have 23 pairs of chromosomes. One pair is known as the *sex chromosomes*, the other 22 pairs being described as *autosomes*. The sex chromosomes are called X and Y. Females have two X chromosomes and males have an X and a Y chromosome. The Y chromosome is much smaller than X and contains only a small number of genes. Genes that are found on the sex chromosomes are said to be sex-linked. For a gene on the X chromosome, a female may be either homozygous or heterozygous. However, a male who inherits only a single X chromosome will have only one copy of the gene. In the case of a recessive gene it is therefore more likely that the trait will be expressed in males. This is the case in red/green colour blindness and for some sex-linked diseases such as haemophilia.

12.9 Genetic fingerprinting

Human beings, with the exception of identical twins, are genetically unique. Genetic fingerprinting takes advantage of this fact to help identify criminals from forensic evidence and for paternity testing.

Genetic fingerprinting makes use of the fact that our DNA contains repetitive DNA sequences – short tandem repeats or STRs. These are present at multiple sites in the DNA and the number of repeats at each site varies between individuals. The technique allows scientists to determine the number of repeats at any given locus in a sample of DNA. By performing this analysis at several loci it is possible to say if a particular sample matches the DNA in another sample with a very high degree of certainty.

12.10 Evolution by natural selection

Each organism inherits a set of characteristics specified by genes acquired from its parents. However, populations can undergo changes in some of these characteristics over time. Such gradual changes are described as the process of *evolution*.

In the late nineteenth century two biologists, Darwin and Wallace, proposed a theory by which such evolutionary change might occur. This was the theory of *evolution by natural selection*. According to Darwin and Wallace, within a population of organisms there is genetic diversity. The alleles of some genes confer characteristics that are better than others in adapting the organism to the particular environment. Those organisms that carry these alleles will, on average, be more likely to survive and reproduce themselves so that these alleles are passed on to the next generation. In this way the population will change over time such that the frequency of such alleles increases. This idea is often referred to as *survival of the fittest* and the environmental factors that favour some genes over others give rise to *selective pressure*. Those organisms that survive are said to have a *selective advantage*. Cumulative changes can eventually give rise to organisms that are genetically so different from the original population that they would be unable to breed with them, and a new *species* has been produced.

Point mutations are one source of diversity that can lead to evolutionary change. Additionally there is evidence that new genes can arise through a process in which existing genes are duplicated, and the two copies of the gene can then diversify though independent mutations. An example of evolution through gene duplication is seen in the pigment genes in the eye that are responsible for colour vision. Colour vision in humans depends on genes for three visual pigments – red, green and blue. In contrast, New World monkeys have only two visual pigment genes – blue and red/green.

Major chromosomal changes can also contribute to evolution. Whereas humans have 23 pairs of chromosomes, chimpanzees and gorillas have 24. It appears that one of the large human chromosomes (chromosome 2) may have derived from two smaller chromosomes that are present in the apes but not in humans.

Evolution usually occurs over long periods of time. In this way all life on earth is envisaged to have evolved from a common ancestor over billions of years. However, where selection pressures are very high, evolutionary change can be observed over much shorter time frames. This is illustrated by the growth in antibiotic resistance that is being observed in a number of bacterial species. As the use of antibiotics has increased, so has the frequency of bacteria with genes conferring resistance to these drugs. One such antibiotic-resistant organism that is currently a cause of extreme concern is MRSA, which stands for methicillin-resistant *Staphylococcus aureus*. *Staphylococcus aureus* is a bacterium that can cause a range of problems from superficial abscesses to life-threatening conditions such as pneumonia and septicaemia. The antibiotic of choice for treating *Staphylococcus aureus* infections used to be penicillin. However, strains started to appear that were resistant to this drug and so in 1960 a new antibiotic, methicillin, was introduced. Within a very short space of time methicillin-resistant strains started to appear and MRSA is now a significant cause of hospital-acquired infections that are very difficult to treat. Fortunately there is an alternative antibiotic that can be used called vancomycin. Worryingly (but not surprisingly), strains of *Staphylococcus aureus* have now started to appear which are also vancomycin-resistant.

> ### CLINICAL EXAMPLE: SICKLE CELL ANAEMIA
>
> Sickle cell anaemia is an inherited disease caused by a mutation in the haemoglobin gene. The mutation, which leads to a single amino acid substitution, when present in the homozygous form causes red blood cells to 'sickle' or change shape at low oxygen concentrations, which leads to blockage of blood vessels. The mutation is autosomal recessive – homozygous individuals have the disease whereas heterozygous individuals are carriers.
>
> Interestingly, the mutation is found in higher frequencies in populations that originate from areas where malaria is endemic. The reason for this is that heterozygous carriers of the sickle cell mutation show an increased resistance to the malaria parasite, which infects red blood cells. This is an important lesson for us – it may seem desirable to eliminate from the population alleles that appear to be deleterious, but we should remember the sickle cell story and the advantage conferred by this disease-associated gene.

12.11 Test yourself

The answers are given on p. 176.

Question 12.5
How is an individual described who has two identical alleles at a particular locus?

Question 12.2
What type of mutation results in the addition of nucleotides to a coding sequence?

Question 12.3
What can be deduced about two alleles that do *not* show independent assortment?

Question 12.4
Which sex chromosomes are present in: (a) human females; (b) human males?

Question 12.5
What is the term used to describe the benefit conferred on individuals by a genotype that enables them to cope better with environmental challenges?

⬡13 Genetic disease

> **BASIC CONCEPTS:**
> A number of diseases are caused by genetic mutations, though diseases
> caused by a single gene are generally rare. Different patterns of inheritance
> are seen depending upon whether the disease allele is dominant or
> recessive and whether it is located on an autosome, a sex chromosome, or
> on mitochondrial DNA. Cystic fibrosis and Huntington disease are both
> caused by genes that are present on autosomes. In cystic fibrosis the
> disease allele is recessive and so the disease will only develop if a child
> inherits a defective allele from both father and mother. The allele that
> causes Huntington disease is dominant and so the disease will develop in
> anyone who has a single copy of the faulty allele. Sex-linked recessive
> diseases such as haemophilia A are more common in boys and can skip a
> generation because the affected father does not pass on his X chromosome
> to his sons. Many common diseases such as asthma are determined by a
> combination of many different genes and environmental factors.

13.1 Introduction

Our genetic make-up has a major role in determining the diseases to which we
are susceptible. Some genes exert a protective effect against particular
diseases, whilst others make us more likely to succumb. The clearest examples
of this are seen in conditions that are caused by a single faulty gene. These
diseases are often serious and sometimes life-threatening, although generally
rare, such as sickle cell anaemia (see Section 12.10). In other more common
diseases, multiple genes may combine, often together with environmental
factors, to determine the likelihood that a person will develop the disease.
Whilst the majority of genetic diseases arise from faulty genes in the nuclear
DNA, a small number relate to genes in the mitochondrial DNA and, because
mitochondria derive from the female gamete, these conditions are inherited
from the mother, such as in mitochondrial myopathy (see Section 6.2).

13.2 Cystic fibrosis: an autosomal recessive disease

Cystic fibrosis is a disease that affects a number of organs and systems but
primarily the lungs and digestive system. Patients with cystic fibrosis usually

start to show symptoms of the disease in the first few months of life. They produce a thick and sticky mucus, which can cause problems with breathing, lead to recurrent respiratory infections, and block the release of digestive enzymes from the pancreas so that food is not properly digested. As a result, children with the condition often fail to gain weight.

The disease is caused by a faulty gene for a membrane transporter protein called cystic fibrosis transmembrane conductance regulator, or CFTR. The *CFTR* gene product is involved in the transport of chloride ions across cell membranes. A large number of different mutations in the *CFTR* gene have been described, but the most common involves a deletion of three nucleotides so that the protein is non-functional because it lacks the phenylalanine amino acid residue normally found at position 508. The *CFTR* gene is encoded on chromosome 7 and the disease allele is recessive, meaning that people with a single copy of the faulty gene are healthy carriers of the disease. In the UK it is estimated that around 1 person in 25 is a healthy carrier. If two healthy carriers have children there is a one in four chance that a child will have the disease (if they inherit a faulty gene from both parents), and a one in two chance that a child will be a healthy carrier (Fig. 13.1).

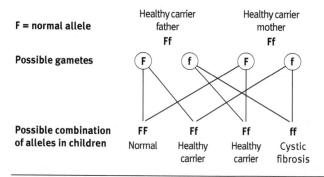

Figure 13.1. Possible allele combinations in children where both parents are healthy carriers of the cystic fibrosis allele (f).

13.3 Huntington disease: an autosomal dominant disease

Huntington disease is an inherited condition in which there is progressive damage to neurons in the brain. The symptoms usually appear in adulthood, and include involuntary movements and impaired mental processes. The disease arises due to a faulty gene on chromosome 4 that encodes the Huntingtin protein. The Huntingtin mutation results in the addition of a number of extra copies of the amino acid glutamine into the protein. Only one copy of the mutant gene is needed to cause the disease. This means that if one parent has the disease and is heterozygous for the gene there is a 50% chance of any children inheriting the condition (Fig. 13.2).

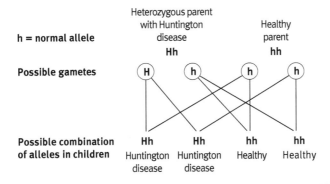

Figure 13.2. Possible allele combinations in children where one parent is heterozygous for the Huntington allele (H).

13.4 Haemophilia A: an X-linked recessive disease

Haemophilia is a group of conditions in which the blood is unable to clot properly, and so patients suffer from repeated bleeding. The usual clotting cascade involves multiple proteins and so a number of different genetic deficiencies can cause haemophilia. In haemophilia A it is the gene for the protein Factor VIII that is defective. This gene is on the X chromosome and the condition is recessive. For this reason it is most frequently seen in males who have only a single X chromosome and so will inherit the disease if they have a single copy of the defective gene. Females who have the defective gene on one of their two X chromosomes are described as carriers of the disease. If a female carrier has children (and the father does not have the haemophilia gene) there is a 50% chance of male children inheriting the defective gene and developing the condition, and a 50% chance of daughters being carriers (Fig. 13.3). If a male with haemophilia has children with a mother who is not a carrier, all of the daughters will be carriers. Note that the father cannot pass on the defective gene to his sons, because they inherit their single X chromosome from their mother.

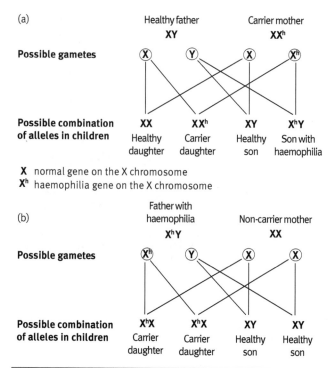

Figure 13.3. Possible allele combinations in children from (a) a healthy father and a carrier mother, and (b) a father with haemophilia and a non-carrier mother.

13.5 Asthma: a disease caused by multiple genes

Asthma is an inflammatory condition affecting the airways. During asthmatic attacks, inflammation in the airways leads to the contraction of smooth muscle and swelling of the mucosa so that the airways become narrow, making it difficult to breathe. Asthmatic attacks may be a result of an allergic response to substances in the environment, or *allergens* (see Section 25.4). Asthma tends to run in families and a number of genes have been identified that appear to predispose people to develop the disease. These include a gene for a protein called filaggrin that is involved in the epithelial barrier (mutations of which may make it easier for the allergens to get into the underlying tissues), a gene for a membrane protease that is expressed in the lung and on smooth muscle tissue, and a number of genes that affect immune responses and inflammation.

Like many diseases caused by multiple genes, environmental factors also influence an individual's chances of developing asthma, including exposure to allergens and infections in early life. For this reason if one member of a pair of identical twins has asthma there is an increased likelihood of the other twin developing the disease, but in about two-thirds of cases this will not happen.

Some of the genes that predispose to the development of a particular disease may also be associated with other diseases. For example, mutations in the filaggrin gene, in addition to being associated with asthma, have also been shown to be associated with eczema, an inflammatory skin condition. Similarly, genes that affect functioning of the immune system may be associated with multiple autoimmune diseases (see Section 25.4).

13.6 Genetic testing

If an individual is suspected to be suffering from a genetic disease they can be tested to see if they have the particular faulty gene. Because it helps the doctor to diagnose the condition, this is called diagnostic testing. Similarly, if a genetic disease is known to be present in a family, then family members may undergo testing to determine if they are carriers of the gene and therefore at risk of passing it on to their children. This is helpful in X-linked conditions, where the sons of a female carrier have a 50% likelihood of developing the disease. It is also helpful in autosomal recessive conditions such as cystic fibrosis where testing of both parents can reveal whether there is a risk of any children inheriting the disease. For diseases that manifest themselves in adulthood, such as Huntington disease, predictive genetic testing is an option for people who are at risk of developing the condition. Whilst this can help with planning for the future, including whether or not to have children, there is presently no cure or preventative treatment and so the benefits have to be weighed carefully against the likely psychological impact of a positive test.

13.7 Gene therapy and genetic diseases

For diseases that are caused by a single, recessive gene, insertion of a healthy copy of the gene into the affected tissues might provide a possible treatment option for the future. This is an approach that is undergoing trials for cystic fibrosis, where the aim is to introduce a healthy copy of the *CFTR* gene into cells in the airways of patients with cystic fibrosis. There are a number of difficulties in this approach, for example, how to incorporate the gene safely, efficiently and in a stable manner into the relevant cells; the fact that the cells are undergoing constant renewal of course means that the therapy would be short-lived and require repeated doses. Gene therapy has shown some promising results in patients with defects of the immune system that are caused by a single gene. Immune cells are produced by the stem cells that give rise to all blood cells, and it is possible to extract these stem cells from a patient, introduce the new gene into the cells, and then put the cells back into the patient. Because the treated cells are stem cells they will give rise to large numbers of healthy immune cells.

In conditions such as Huntington disease, which are caused by a dominant gene, researchers are exploring the possibility of 'silencing' the disease gene using an approach called RNA interference. This involves introducing into the affected cells RNA molecules that will bind to the mRNA for the Huntingtin protein and interfere with its production. As with gene therapy, one of the major challenges is to find ways of efficiently incorporating the interfering RNA molecules into the cells.

13.8 Test yourself

The answers are given on p. 176.

Question 13.1
Diseases that are caused by mitochondrial genes can be passed on by the daughters, but not the sons, of a person with the disease. Why?

Question 13.2
What is the function of the protein encoded by the gene involved in cystic fibrosis?

Question 13.3
For a genetic disease that shows an autosomal dominant pattern of inheritance all children who inherit the disease allele will be affected. True or false.

Question 13.4
A number of different genes can cause haemophilia. Why?

Question 13.5
What is the probability that a man with an X-linked disease will pass the condition on to his sons?

14 Epithelial tissues

> **BASIC CONCEPTS:**
> Epithelial cells form the covering of the external and the internal surfaces of the body. They also form glands. They provide the interface between the organs of the body and their external and internal environments. Epithelia are vital in ensuring that the internal environment is maintained in a state consistent with life. They perform a range of important transport functions. Understanding the role of epithelia is vital to an understanding of the body's interaction with its environment.

14.1 Introduction

Collections of cells that are specialised to perform a specific function are called *tissues*. The organ systems of the body are made up of four basic tissues – *epithelia*, *connective tissue*, *muscle* and *nervous tissue*.

An *epithelium* (plural *epithelia*) is a continuous layer of cells covering an internal or external body surface. The *epidermis* of the skin, which can be up to 200 cells thick, is an epithelium, as is the delicate single-cell lining of the alveoli of the lungs. Glands, which are the secretory organs of the body, are also derived from epithelial tissue.

14.2 Classification

Epithelia can be classified as *simple* or *stratified*. A simple epithelium consists of one layer of cells and a stratified epithelium has more than one layer of cells. In addition, epithelia are further described according to the shape of the cells in their upper layer. Three basic shapes are described – *squamous* (flattened), *cuboidal* and *columnar* (Fig. 14.1).

Epithelia have many functions and these are reflected in their cellular organisation as is summarised in the table below.

Classification	Function	Location
Simple squamous	Facilitate diffusion and transport from one compartment of the body to another	Alveoli of lungs Capillaries Kidney glomerulus
Simple cuboidal	Often associated with transport of ions across cells	Kidney tubule Thyroid follicle
Simple columnar	Secretion and absorption	Small and large intestine Pancreas
Stratified cuboidal	Lining of ducts	Sweat gland duct
Stratified squamous	Protection against wear and tear	Skin Vagina Oral cavity

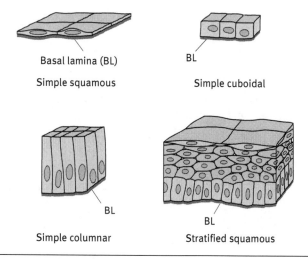

Basal lamina (BL)

Simple squamous

BL

Simple cuboidal

BL

Simple columnar

BL

Stratified squamous

Figure 14.1. Types of epithelia

14.3 Adhesion

It is important that epithelial cells maintain a coherent sheet-like structure by being able to adhere both to neighbouring cells and to the underlying tissues. The majority of epithelial cells sit on a thin protein sheet known as the *basal lamina*. The basal lamina is secreted by the epithelial cells themselves and is a fine meshwork of proteins, including collagen (see Chapter 2). This extracellular meshwork provides attachment both for the epithelial cells above and for the connective tissue below and is important in linking epithelial sheets to the underlying connective tissues.

The interactions between adjacent epithelial cells depend on three main types of intercellular junctions.

- *Desmosomes* are adhesions between adjacent cells characterised by dense circular protein plaques inside the cell membrane. These act as an anchor point for extracellular protein bridges and as a point of attachment within the cell for cytoskeletal proteins. Desmosomes also link the epithelial cells to the basal lamina and thus anchor the epithelial sheet in place.
- *Tight junctions* are areas where the cell membranes of adjacent cells are held close together by shared trans-membrane proteins. They act to restrict the movement of ions and other substances between epithelial cells and thus help to prevent epithelia being 'leaky'.
- *Gap junctions* are small pores formed between adjacent cells that permit the free movement of water, ions and small molecules such as glucose and amino acids. Gap junctions allow cells in an epithelium to act together in response to external stimuli. In stratified epithelia gap junctions allow the passage of metabolic substrates to cells that are not in direct contact with the blood supply. Gap junctions are also found in excitable tissues such as cardiac muscle (see Chapter 16).

CLINICAL EXAMPLE: BULLOUS PEMPHIGUS

This is a rare autoimmune disease in which the body develops antibodies against proteins of the desmosomes in skin. This results in the epidermis of the skin lifting away from the underlying connective tissue and leads to the formation of blisters. The disease can be life-threatening.

14.4   Test yourself

The answers are given on p. 176.

Question 14.1
How many layers of cells occur in a simple epithelium?

Question 14.2
How are epithelia containing multiple layers of flattened cells described?

Question 14.3
What is the term used to describe the protein sheet that underlies an epithelium?

Question 14.4
What type of junctions in epithelia are characterised by shared membrane proteins between adjacent cells?

Question 14.5
What type of junctions link an epithelium to its basal lamina?

15 Connective tissues

> **BASIC CONCEPTS:**
> Connective tissues, such as bone and cartilage, are key components of the skeleton. As such they are essential in locomotion. Other connective tissues are important structural elements of many organs and some allow the body to store energy in the form of fat. Connective tissues are all composed of the same basic elements but vary in how these are put together. Understanding the structure–function relationships in connective tissues is the key to understanding their diverse roles.

15.1 Introduction

There are a wide variety of types of *connective tissue* in the body – including adipose tissue, tendon, cartilage and bone – as well as the specialised tissues of the teeth – enamel and dentine. All connective tissues have the same basic elements in that they consist of specialised cells surrounded by an *extracellular matrix*. The matrix is made up of a hydrated glycosaminoglycan gel in which are embedded protein fibres. The matrix is secreted and maintained by the cells which sit within it. In mineralised connective tissues, such as bone, the matrix also contains crystals of calcium phosphate which provide rigidity.

15.2 Glycosaminoglycans

These are unbranched polysaccharide chains with repeating disaccharide units. There are a number of different *glycosaminoglycans* (GAGs) depending on the composition of the disaccharide. Some GAGs are also sulphated. Individual GAGs are assembled onto a protein core (like the teeth of a comb) to form a larger structure known as a *proteoglycan*. These proteoglycans are in turn linked together to form a supermolecule known as an *aggrecan*. All GAGs carry significant negative charge which draws osmotically-active cations (predominantly Na^+) into the matrix which in turn attract water and create a hydrated gel. The length and composition of the individual GAGs within the aggrecan molecule will determine the total negative charge and thus the amount of water drawn in. This will determine the turgidity (degree of swelling) of the extracellular matrix. A connective tissue such as cartilage has a high water content and is thus able to resist compression.

15.3 Fibres

The fibrous components of the connective tissue matrix comprise principally collagen and elastin.

Collagen

Collagen provides tensile strength and flexibility. The presence of parallel bundles of collagen fibres in a tendon provides a good example of these properties. As discussed in Chapter 2, collagen is a fibrous protein assembled from three helical α-chains. There are up to 25 collagen α-chains coded for in the genome and the collagen type is determined by the precise combination of the α-chains from which it is made. The collagen type which is found in most major connective tissues, with the exception of cartilage, is Type 1 and this is made up of two α1 and one α2 chains. Once the individual triple helical collagen molecules have been secreted into the extracellular matrix they begin to self-assemble into fibrils and then into fibres. A collagen fibre can be up to 10 μm in diameter. Not all collagen types assemble into fibres – for example, Type IV collagen, which is found in the basal lamina, assembles as a fine meshwork.

> ### CLINICAL EXAMPLE: SCURVY
>
> Scurvy is a condition once associated with sailors on long voyages and is characterised by weakness, bleeding under the skin, teeth falling out and ultimately death. It is caused by a lack of fresh fruit leading to vitamin C deficiency which results in collagen fibre assembly malfunction. The reason for this is that the assembly of collagen molecules into fibres depends on the presence of the amino acid 4-hydroxyproline in the collagen molecule. The production of this hydroxylated amino acid depends on the presence of vitamin C.

Elastic fibres

Elastic fibres have a composite structure consisting of a core of a coiled and cross-linked protein, elastin, surrounded by a fine meshwork of another protein, fibrillin. Elastic fibres allow connective tissues to be distorted and then to return to their original shape. The walls of blood vessels contain elastic fibres which allow them to expand and then recoil as the blood pulses through them (see Chapter 20).

Other matrix proteins

In addition to collagen and elastin, the extracellular matrix of connective tissues contains a wide variety of glycoproteins with diverse functions, which include regulating the assembly of the other components of the matrix, acting as binding sites for adhesion molecules on cell surfaces, acting as molecular markers for cell migration pathways and inhibiting unwanted calcification.

15.4 Cells

A range of cell types are found within connective tissues. The cells responsible for the secretion of the extracellular matrix are known as '...blast' cells and are further described according to the nature of the connective tissue they produce. Thus *fibroblasts* secrete fibrous connective tissues such as tendon and ligament while *chondroblasts* lay down cartilage and *osteoblasts* deposit bone.

Once these cells have deposited a matrix around themselves they become the mature cells of the connective tissue and are responsible for the subsequent repair and maintenance of the matrix. Death of these cells would result in resorption of the matrix. The mature cells are known as '...cytes' – as in *fibrocytes*, *adipocytes* (in adipose tissue), *chondrocytes* and *osteocytes*. In fibrous connective tissues both fibrocytes and stem cells within the connective tissue retain the capacity to divide and will do so in response to injury or insult. This can lead to scarring (*fibrosis*) if excessive cellular activity with consequent deposition of collagen fibres into the extracellular matrix occurs.

Connective tissue matrices also play host to many other cell types:

- *mast cells* – which secrete heparin and histamine and are important in producing localised inflammation in response to injury and in certain types of allergic reaction;
- *plasma cells* – which are responsible for antibody production;
- *macrophages* – which are blood-derived scavenger cells.

15.5 Test yourself

The answers are given on p. 176.

Question 15.1
What substance of the extracellular matrix is responsible for drawing in water to form a gel?

Question 15.2
What protein provides strength and flexibility to the extracellular matrix?

Question 15.3
What name is given to the cells responsible for laying down cartilage?

Question 15.4
What are the mature cells in cartilage called?

Question 15.5
What is the function of mast cells in connective tissues?

(16) Excitable tissues

BASIC CONCEPTS:
Muscle and nerve are known as excitable tissues. This is because in response to stimuli they can rapidly alter the disposition of ions across their membranes, leading to changes in electrical potential. These electrical changes can be transmitted along the membranes of excitable cells. In muscle cells this will trigger contraction. Nerve cells use electrical changes to transmit information and are capable of exciting adjacent nerve cells to generate a network in which cooperative activity can lead to higher cognitive processes such as thought. Understanding the basic processes by which electrical changes are generated is important in understanding the basis of thought, senses and movement.

16.1 Introduction

Muscle and nerve are both described as *excitable tissues*. This relates to the fact that both types of tissue have membranes which maintain a significant electric potential difference between the inside and outside of the cell. This is normally expressed in millivolts, with the outside of the cell considered as being at 0 mV. The inside of the cell is negative with respect to the outside and in excitable tissues normally lies between −60 and −80 mV. This potential difference (known as the *resting membrane potential*) can be altered rapidly by the movement of ions across the membrane, and when this results in the inside of the cell becoming less negative is known as *depolarisation* or *excitation*.

16.2 Membrane potential

In order to understand how excitable tissues work it is useful to understand how the potential difference across a cell membrane is generated. In order to explain this we will use a hypothetical situation in which we consider only a limited number of the charged molecules which are found within cells and in the extracellular fluid.

As explained in Chapter 1, the extracellular fluid (ECF) has high levels of Na^+ ions and the intracellular fluid (ICF) has high levels of K^+ ions. Consider a system at equilibrium in which the ECF and ICF contain equal numbers of ions and have equal numbers of positive and negative charges – and thus are electrically neutral. In this situation there is no difference in charge across the membrane. If we were now to open in the membrane an ion channel selective for K$^+$ ions then the movement of this ion through the channel would depend on two forces. The first would be a force generated by the concentration gradient for K$^+$ ions across the membrane where internally the K$^+$ ion is at a high concentration (160 mmol) and externally it is at a low concentration (4 mmol). Thus simple diffusion would tend to cause K$^+$ ions to move out of the cell (see Fig. 16.1). However, every K$^+$ ion that moves out of the cell will create an electrical imbalance across the membrane since a positive ion moving out will leave a spare negative ion behind which cannot cross the membrane because there are no channels open for it to use. The developing difference in charge created by leakage of K$^+$ ions out of the cell will make the movement of further K$^+$ ions increasingly difficult as they are drawn back by the negative charge. The outward force generated by the concentration gradient is opposed by a developing electrical gradient. Ultimately these two forces will reach equilibrium and a membrane potential will be established at which net inward and outward movement of K$^+$ ions is balanced.

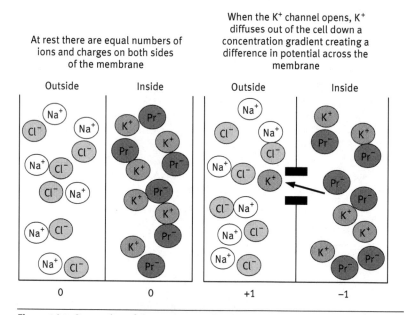

Figure 16.1. Generation of the resting membrane potential

For a single ion, like K$^+$, this is known as the *equilibrium potential* (E_k) and is described by an adaptation of the Nernst equation (below) in which R is the gas constant, F is the Faraday constant, T is the temperature (°Kelvin), Z is the valence (charge) of the ion and the terms in square brackets are the concentrations of K$^+$ in the ECF, [K$^+$]$_e$ and the ICF, [K$^+$]$_i$:

$$E_k = \frac{RT}{ZF} \log \frac{[K^+]_e}{[K^+]_i}$$

When this equation is applied to the physiological values for K^+ it gives an answer of −75 mV. That this value is remarkably close to the actual resting membrane potential of excitable tissues is because the only major ion which is free to move at rest in these cells is K^+ for which there exists a specific leak channel. Movement of the other ions contributes very little to the resting membrane potential.

16.3 Muscle

Muscle cells are capable of contraction either spontaneously or in response to external stimuli. There are three main types of muscle in the body – *skeletal muscle*, *cardiac muscle* and *smooth muscle*.

Skeletal muscle

This is the type of muscle which is found in the musculoskeletal system and is responsible for voluntary movement. An individual skeletal muscle (such as the biceps in the arm) is attached at both ends to a bone via a *tendon* (see Chapter 26). Contraction of the muscle results in the movement of one bone relative to the other. The cells of skeletal muscle are elongated, multinucleated tubular structures known as *myocytes* or muscle fibres. Each muscle fibre is innervated centrally, in an area known as the neuromuscular junction, by the terminal branch of the axon of a motor neuron whose cell body is located in the central nervous system.

Internally, the cytoplasm of the muscle fibre is filled with structures known as *myofibrils* which consist of repeated units known as *sarcomeres*. In turn, sarcomeres are composed of microfilaments of the proteins *actin* and *myosin* and it is the interaction of these proteins which is responsible for contraction (see Fig. 16.2). Each sarcomere contains a central bundle of myosin filaments which are overlapped at both ends by an array of actin filaments. Each myosin filament is surrounded by, and can interact with, six actin filaments. The myosin filament is assembled from a number of myosin molecules, each of which contains a rod-like domain and a flexible head which is capable of binding actin and breaking down ATP. The actin filaments at both ends of the sarcomere insert into a protein band known as the *Z-line*. Thus a sarcomere runs from one Z-line to another. The part of the sarcomere containing the thicker myosin filaments tends to stain more darkly when viewed down the microscope and is known as the *A-band*. The area of the sarcomere containing only actin is known as the *I-band*. The alternating dark and light bands give skeletal muscle a striped appearance – hence it is sometimes referred to as *striated muscle*.

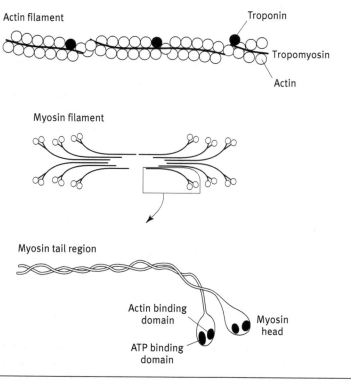

Figure 16.2. Structure of actin and myosin filaments

The process of muscle contraction depends on the interaction between myosin and actin filaments. Basically, the extent to which the actin and myosin filaments overlap is increased during contraction and this results in the shortening of the sarcomere. The mechanism by which this occurs is outlined below and in Fig. 16.3.

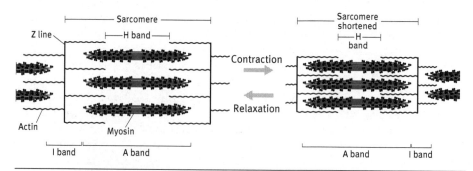

Figure 16.3. Muscle contraction

Nerve impulses arriving at the neuromuscular junction trigger the release of the neurotransmitter acetylcholine (ACh). Binding of ACh to specific receptors on the muscle membrane causes the opening of Na^+ channels, leading to the influx of Na$^+$ ions and depolarisation of the muscle cell at the neuromuscular junction.

The depolarisation spreads away from the neuromuscular junction, across the surface of the muscle fibre as voltage-dependent Na$^+$ channels adjacent to the initial site of depolarisation are triggered to open. The depolarisation is also transmitted through a system of invaginations of the cell membrane (*t-tubules*) which run into the centre of the muscle fibre. The t-tubules are closely related to expansions of smooth endoplasmic reticulum (SER) within the cytoplasm of the muscle fibre. The depolarisation of the t-tubules triggers the opening of voltage-dependent Ca^{2+} channels in the membrane of the SER allowing Ca^{2+} ions to enter the cytoplasm of the muscle fibre.

The Ca^{2+} ions bind to a protein known as troponin which, in conjunction with a second protein tropomyosin, lies along the actin filament and blocks access to the myosin binding site on the actin molecules. Ca^{2+} binding to troponin induces a conformational change which exposes the myosin binding sites. The myosin heads then cross-link with the actin filament.

The myosin acts as an ATPase and breaks down ATP to ADP. The energy from this reaction is used to cause a change in the angle of the myosin head such that the actin filament is drawn towards the centre of the A-band.

The myosin releases the actin and replaces the ADP it now has bound to its head with ATP. The angle of the head changes back to its resting conformation.

If Ca^{2+} ions are still present then these events are repeated and contraction proceeds. Contraction is terminated by the active removal of Ca^{2+} ions from the cytoplasm of the muscle fibre back into the SER. This occurs when depolarisation at the neuromuscular junction has been terminated by the breakdown of ACh by the enzyme acetylcholinesterase.

The force of the contraction generated by the shortening of the sarcomeres is transmitted through their attachment to the cell membrane by a set of proteins, including linker proteins known as the *dystrophins*, and from there to the fibres of the connective tissue in which the muscle fibres are embedded. These connective tissue fibres connect with more densely packed fibres of the tendon or ligament at the end of the muscle, and thus force is transmitted to the bone.

CLINICAL EXAMPLE: MUSCULAR DYSTROPHY

Muscular dystrophy is a wasting disease of muscle which affects young boys. It is an X-linked genetic disease caused by a mutation in the gene for a dystrophin. The improper attachment of myofilaments to the cell membrane of muscle fibres results in the membranes of the muscle fibres gradually becoming damaged and the cells degenerating. The effects of the mutation are cumulative and so those affected may have normal muscle function in infancy but become progressively weaker and they normally die in adolescence.

Cardiac muscle

Cardiac muscle is a specialised form of muscle found only in the heart. It consists of a network of branched cells joined end-to-end by a specialised junctional complex known as the *intercalated disc*. This structure consists of alternating segments of desmosomal-like cell adhesions and gap junctions. The gap junctions allow the free movement of ions from one cardiac muscle cell to another, and thus enable cardiac muscle to operate as an electrical *syncytium* in which depolarisation of one cell leads to depolarisation and contraction of all connected cells.

Internally, cardiac muscle cells contain myofibrils similar to those found in skeletal muscle and thus cardiac muscle also has a striated appearance. The mechanism of contraction of cardiac muscle cells is similar to that in skeletal muscle depending on the internal spread of depolarisation through t-tubules and the release of Ca^{2+} ions which bind to troponin. Cardiac muscle cells are able to depolarise spontaneously and this property is used by pacemaker cells in the heart wall to generate rhythmic beating (see Chapter 20).

Smooth muscle

Smooth muscle is normally found in the walls of internal organs – such as the gastrointestinal tract, the bladder and the uterus – and is responsible for slow and sustained involuntary movements. The cells of smooth muscle (Fig. 16.4) are spindle shaped and in most tissues electrical continuity is provided through the presence of connecting gap junctions. Internally the basic unit of contraction is the sarcomere but there are significant differences in terms of both the organisation of the sarcomeres and the mechanism of contraction between smooth muscle and that described above for skeletal and cardiac muscle.

Sarcomeres within smooth muscle cells run diagonally across the muscle cell and are attached at both ends to a structure known as a *dense body* rather than into a Z-line. Contraction of the sarcomere results in a smooth muscle cell becoming shorter and fatter. Most smooth muscle cells are in a permanently semicontracted state and this resting level of contraction is known as *muscle tone*. Smooth muscle contraction can then be regulated either upwards or downwards leading to further contraction or to relaxation.

Dense body   Sarcomere

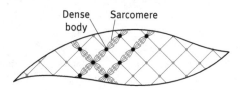

Figure 16.4. Smooth muscle cell

Contraction in smooth muscle cells, as in striated muscle, is regulated by the levels of intracellular calcium. Contraction is initiated by the entry of Ca^{2+} ions into the cytoplasm, either through regulated Ca^{2+} channels in the cell membrane or by release from intracellular stores. The cytoplasmic Ca^{2+} ions bind to the protein *calmodulin* to produce an activated Ca^{2+}-calmodulin complex. This interacts with the inactive cytoplasmic protein *myosin light chain kinase* (MLCK) to convert it into active MLCK. The MLCK phosphorylates a component of the myosin head enabling it to interact with actin. At this point the myosin converts ATP to ADP and the energy released is used to adjust the angle of the myosin head to slide the actin filament towards the centre of the sarcomere. If Ca^{2+} ion levels remain high the contraction may be maintained, or if Ca^{2+} ion levels fall then relaxation may ensue.

16.4 Nerve

The nervous system is made up of two main cell types – *neurons* and *glia*. The neurons are the main excitable cells within the nervous system and the glia act predominantly as supporting cells functioning to myelinate the axons of neurons and to regulate the ionic and nutrient content of the local microenvironment. The nervous system will be considered in detail in Chapter 19 and in this section only the structure of neurons and their basic electrical properties will be considered.

Neurons

A neuron has three main components as shown in Fig. 16.5. There is a *cell body*, which contains the nucleus, from which arise two structures – *dendrites* and an *axon*. Information in the nervous system is represented as patterns of electrical activity within its neurons. The nervous system is constructed to allow the flow of information from one set of neurons to the next, and the neurons are organised into complex networks in which each cell may receive information from up to 10 000 other cells. This transmission of information occurs at specialised structures known as *synapses*, which are formed between the axon of one neuron and the dendrites of the next. The dendrites of a neuron form an extensively branched network arising from the cell body, and the membrane of the dendrites is capable of being depolarised at multiple points as the result of activity at synapses (see below). The depolarisation of the dendrites spreads passively towards the cell body, which acts to sum the activity in all the dendrites which arise from it. Depending on the level of summed activity arriving at the cell body, a wave of depolarisation known as an *action potential* may be initiated, and then transmitted along the axon to its terminal. Arrival of the action potential at the axon terminal triggers release of a chemical neurotransmitter which acts at the synapse to depolarise the adjacent dendritic membrane (see below).

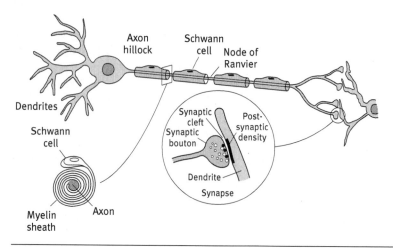

Figure 16.5. A neuron

Action potential

The action potential is a wave of depolarisation which can be transmitted along an axon. It is an all-or-nothing phenomenon, meaning that it has the same amplitude regardless of both the magnitude of the stimulus and wherever it is recorded along an axon. An action potential may be represented graphically as shown in Fig. 16.6. This indicates what would be measured if one recorded the membrane potential at a fixed point along the axon with a recording electrode placed inside the axon. The vertical axis is voltage (the membrane potential) and the horizontal axis is time. At rest (point *A*) the membrane is at a fixed resting membrane potential of −70 mV. As a result of local depolarisation the membrane is gradually depolarised (*B*) until it reaches a threshold point (*C*) of around −50 mV. At this point, large numbers of voltage-sensitive Na^+ channels in the axonal membrane open and Na^+ ions flood into the cytoplasm of the axon down their concentration gradient. This rapid influx begins to depolarise the membrane. Because Na^+ is now the major ion moving across the membrane, the Nernst equation can be applied for Na^+ and the membrane potential heads rapidly (*D*) towards the Na^+ equilibrium potential, which is +50 mV. However, before the membrane can reach the Na^+ equilibrium potential, the depolarisation triggers the opening of K^+ channels in the membrane. This results both in the efflux of K^+ ions and the closure of the Na^+ channels. Now K^+ is the major ion moving across the membrane (*E*) and so the membrane potential heads back towards the K^+ ion equilibrium potential of −75 mV. At point *F* the membrane hyperpolarises (goes below its resting membrane potential) as the K^+ ion equilibrium potential is approached. The K^+ channels then close and the activation of the Na/K ATPase restores the ionic balance across the membrane. During this period of hyperpolarisation there is an initial phase

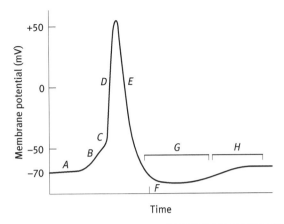

Figure 16.6. Action potential

(*G*) during which the Na⁺ channels are rendered incapable of reopening and thus no further action potentials may be generated in this region of membrane. This is known as the *absolute refractory period*. This is functionally important because action potentials normally travel only away from the cell body. This wave of depolarisation is propagated along the axon by depolarisation in one region triggering the opening of Na⁺ channels in the adjacent membrane regions. Because the membrane which has just been depolarised is refractory, it cannot undergo further depolarisation and thus the action potential can move in only one direction. Following the absolute refractory period there is a further period during which the membrane is slightly hyperpolarised (*H*) and thus further away from threshold value and more difficult to depolarise again. This is known as the *relative refractory period*.

For most neurons, information is coded in the frequency and pattern of action potentials which they transmit.

Myelination

Conduction of an action potential along an axon is a relatively slow process, depending as it does on the sequential opening of Na⁺ channels. In thin unmyelinated axons the rate of conduction can be as slow as 2 m/s. Given that reaction times between receiving a visual cue and responding with a finger movement can be as fast as 0.2 s and that the distance from the brain to the finger tip is at least 1 m, it is clear that most axons are able to conduct action potentials at a much faster rate. Indeed the fastest axons in our body have a conduction velocity of 100 m/s. Conduction rates are enhanced because axons are surrounded by a *myelin sheath* (see Fig. 16.5). This is formed from the cell membrane of a specialised glial cell. In the brain these cells are known as oligodendroglia and in peripheral nerves myelination is carried out by *Schwann cells*. A myelin sheath consists of a number of concentric layers of the cell membrane of the glial cell wrapped around the axon. The membranes within the myelin sheath contain specialised lipids and associated proteins such as myelin basic protein. A basal lamina lies between the myelin sheath and the axonal membrane.

Because it is predominantly lipid, the myelin sheath acts as an electrical insulator. Along the length of an axon the myelin sheath is not continuous and at the points between adjacent glial cells the axonal membrane is uncovered at areas known as the *nodes of Ranvier*. These nodes are the only areas of the axon where depolarisation may occur. During transmission of an action potential along a myelinated axon, depolarisation at one node triggers immediate depolarisation at the next node and the action potential leaps from node to node down the axon. This is known as *saltatory conduction* and greatly enhances both the speed and efficiency of conduction by reducing the number of channel opening events required to get from one end of the axon to the other.

CLINICAL EXAMPLE: MULTIPLE SCLEROSIS

MS is a disease in which the myelin sheath of axons is destroyed. It is caused by the immune system attacking the myelin sheath by directing an immune response against components such as myelin basic protein. The resultant demyelination produces significant impairment of function in the affected pathways. Individuals with MS may suffer from loss of vision, loss of motor function or loss of peripheral sensation. The disease is particularly devastating because of its random nature – there is no way of knowing when or where it will strike and which functions will be affected. At present there is no cure and the only treatments are drugs which suppress immune function, such as corticosteroids.

Synaptic transmission

Although each neuronal cell body gives rise to only a single axon this may branch many times at its termination and thus make contact with many dendrites. The contact between an axon terminal and a dendrite is known as a *synapse* (see Fig. 16.5). There is a small swelling at the end of the axon which is filled with many vesicles containing a chemical neurotransmitter. When the action potential arrives at the end of the axon it triggers the opening of voltage-dependent Ca^{2+} channels and the subsequent influx of Ca^{2+} ions triggers vesicle fusion with the presynaptic membrane (Fig. 16.6). Each vesicle then releases into the synaptic cleft a small amount of neurotransmitter known as a quantum. Vesicle recycling occurs in that once a vesicle has released its contents it is internalised back into the axon terminal and refilled with neurotransmitter. On the dendritic side of the synaptic cleft are membrane receptors that bind the neurotransmitter. Some neurotransmitters, such as acetylcholine, are known as excitatory and open

Na^+ channels in the postsynaptic membrane. The effect of this is to produce small depolarisations. Other neurotransmitters, such as gamma-aminobutyric acid (GABA), are inhibitory and these open Cl^- channels, tending to oppose depolarisation. The net effect of these events at all the synapses contacting one neuron may lead to the generation of an action potential at the cell body. Termination of the interaction between the transmitter and the receptor is achieved either by enzymatic breakdown of the transmitter in the synaptic cleft or by the re-uptake of the transmitter into the axon and its internal recycling.

CLINICAL EXAMPLE: DEPRESSION AND SSRIS

Serotonin is a neurotransmitter which acts to modify the activity of neurons in the brain, which are important for our control of mood. It is known that one of the causes of depression is a reduced level of serotonin within certain areas of the brain. A class of antidepressant drugs known as selective serotonin re-uptake inhibitors (SSRIs) block the mechanism which takes serotonin back up into the axon terminal and thus increases the amount and duration of action of the serotonin present in the synaptic cleft. This helps to alleviate the symptoms of depression.

16.5   Test yourself

The answers are given on p. 176.

Question 16.1
Which cation, found in higher concentration in the cytoplasm than in the extracellular fluid, is principally responsible for maintaining the potential difference across a cell membrane?

Question 16.2
What two proteins present in the microfilaments of muscle cells are mainly responsible for muscle contractions?

Question 16.3
What neurotransmitter released at the neuromuscular junction causes the depolarisation of the muscle cell membrane?

Question 16.4
What are the three main components of a neuron?

Question 16.5
What is the function of the myelin sheath?

⬡17 Homeostasis

> **BASIC CONCEPTS:**
> Homeostasis is the regulatory process by which the body maintains a constant internal environment. For each parameter to be regulated it is important to have a sensor mechanism which allows its current status to be monitored. The current status is then checked against a predetermined 'ideal state' and deviations from this will initiate a response that will move the measured parameter back towards the 'ideal'. The response may be achieved via hormonal or neural mechanisms. An understanding of homeostasis underpins an understanding of much of physiology.

17.1 Introduction

Homeostasis is the set of processes by which the body strives to maintain a constant internal environment. This principally involves regulation of the composition of the extracellular fluid compartment. The key properties that are regulated are *temperature*, *blood pressure* and *volume*, *plasma metabolite levels*, *plasma oxygen levels* and *plasma osmolarity*. A constant internal environment is needed to ensure that cells can function normally. For example, if plasma potassium levels are not maintained within very tight limits then membrane potentials in excitable tissues are affected. This can rapidly result in heart failure and death.

Most homeostatic mechanisms in the body are based around a feedback loop in which the levels of the substance or parameter to be regulated are constantly monitored. Fluctuations of these levels away from a preset norm are detected and signals passed to an effector mechanism which adjusts the levels back into the normal range. The signalling mechanisms for both the feedback and effector arms of the system can be either hormonal or neural. Once the levels are back within the normal range the effector mechanism is shut down.

In many systems, constant levels are achieved by the action of two opposing mechanisms – one to elevate the parameter and one to depress it. Homeostasis often results in oscillations of any given parameter about the set level because feedback results in constant adjustments up and down. An effective homeostatic mechanism will minimise these fluctuations.

17.2 Regulation of plasma glucose

Homeostasis of plasma glucose offers an example of a mechanism which relies predominantly on hormonal feedback loops (Fig. 17.1). Following a meal, absorption of glucose from the small intestine results in a rapid rise of plasma glucose. Prolonged elevation of plasma glucose can harm the body in a number of ways, including damage to small blood vessels in the retina and kidney. It is thus important that plasma glucose levels are maintained within a normal range of about 4.5–5.6 mmol/l.

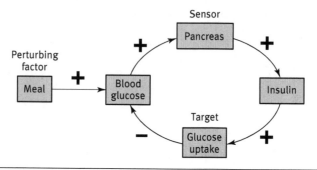

Figure 17.1. Feedback loop in the control of blood glucose

Detection of plasma glucose levels is achieved by beta cells located within the pancreas gland. The beta cells contain vesicles which store the protein hormone insulin. Glucose enters the beta cells and triggers the release of the insulin into the bloodstream. The release of insulin is stimulated at plasma glucose levels above 5 mmol/l. Insulin travels in the bloodstream to its target tissues, which are primarily the liver and skeletal muscle. Insulin binds to its receptor and enhances the uptake of glucose into the cells of these tissues. Once plasma glucose levels have fallen to within the normal range then insulin release is shut down and glucose uptake is down-regulated.

If plasma glucose levels fall below the normal range then a second type of cell in the pancreas, the alpha cell, is stimulated to produce the hormone glucagon. This acts on liver to stimulate the breakdown of glycogen into glucose which is then released into the bloodstream.

> **CLINICAL EXAMPLE: DIABETES**
>
> Diabetes is characterised by a loss of control over plasma glucose resulting in increased levels of glucose in the blood (hyperglycaemia).
>
> In Type 1 diabetes, which particularly affects adolescents, the body's own immune system destroys the beta cells in the pancreas. Thus the body is no longer able to respond to increases in plasma glucose by secreting insulin, and meals are followed by rapid rises in plasma glucose, the osmotic and metabolic effects of which can lead to coma.
>
> In Type II diabetes, which tends to affect older obese patients, the target cells in liver and muscle lose their ability to respond to insulin. This is the result of a desensitisation of insulin receptors. Patients with Type II diabetes show chronically elevated plasma glucose levels. In the long term this can damage small blood vessels leading to retinal degeneration and blindness and to loss of blood flow to the legs and feet causing death of tissues in the feet and their eventual amputation.

17.3 Thermoregulation

Maintenance of a constant internal temperature is vital for the correct functioning of body systems. If body temperature rises above about 41°C then proteins can start to denature and metabolic systems break down. Below 32°C metabolic processes slow down so much that energy production is negligible and cell functions are seriously impaired. *Thermoregulation* is primarily achieved through the actions of the autonomic nervous system (see Chapter 19) and provides an example of a neural homeostatic mechanism.

Neurons in a region of the brain called the hypothalamus (see Fig. 17.2) act as the primary temperature regulatory centre. They monitor core temperature by being sensitive to fluctuations in the temperature of the blood which flows through the region. They also monitor peripheral temperature by receiving neural input from thermoreceptors in the skin, which respond to both cold and warmth. Within the hypothalamus there is a set of cells which increase their firing rate in response to a drop in temperature and another set which respond to a rise in temperature. In response to a change in temperature, there are a number of mechanisms (neural, metabolic and behavioural) which the body can utilise to alter heat generation and to regulate heat loss at the periphery.

As temperature falls, an increase in the activity of the sympathetic nervous system (see Chapter 19) leads to constriction (narrowing) of blood vessels near the body surface, causing a diversion of blood flow away from the vessels immediately underneath the epidermis of the skin – thus reducing heat loss from the surface. In addition, the muscles which are attached to hair follicles are stimulated to contract, causing hair to stand on end. In furry

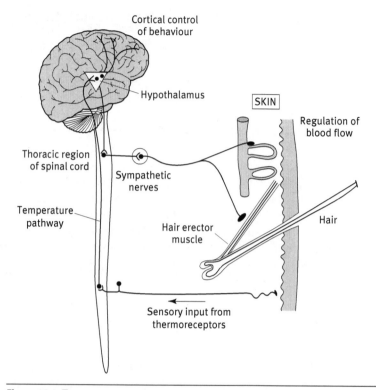

Figure 17.2. Temperature regulation

animals this leads to an increase in the thickness of the layer of air trapped in the hair – thus improving insulation. The principal effect in humans is to produce 'goosebumps'. Skeletal muscles are stimulated to contract and relax repeatedly, causing shivering. This increases their heat production as a result of increased metabolic activity. The thyroid and adrenal glands (see Chapter 18) are stimulated to release thyroid hormone and adrenaline respectively. Both of these hormones act to increase metabolic activity in organs such as the liver and thus generate heat. In addition, behavioural programmes are initiated which lead to simple behaviours such as huddling up and to complex adaptive behaviours such as seeking shelter and putting on additional clothing.

Newborn babies have an additional thermoregulatory mechanism in that they possess a specialised adipose tissue known as brown fat.

Activation of this tissue leads to the expression within mitochondria of a protein which uncouples the products of the citric acid cycle from the electron transport chain (see Chapter 8) with the result that the energy normally used to generate ATP is converted directly to heat.

As temperature rises, peripheral blood vessels dilate, resulting in an increase in blood flow near to the skin and consequent heat loss through radiation. Activation of sweat glands causes an increased secretion of sweat onto the skin surface where evaporation causes further heat loss. Activation of the muscles associated with hair follicles is reduced and so hairs will lie flat and close to the skin surface, encouraging heat loss through convection and evaporation. The production of thyroid hormone and adrenaline is also reduced with the resultant slowing of metabolic rates. Heat-reducing behaviours such as shade seeking and clothes shedding are induced.

CLINICAL EXAMPLE: FEVER

Fever is a response by the body to help fight infection. The raised temperature improves the performance of the cells of the immune system and impairs the ability of microorganisms to replicate. Toxins from the microorganisms induce the production of a protein called IL-1 from cells of the immune system. IL-1 acts within the blood vessels of the hypothalamus to stimulate production of prostaglandin E2, which acts to reduce the firing of warm sensitive neurons and to initiate mechanisms which will lead to a rise in body temperature.

17.4 Test yourself

The answers are given on p. 176.

Question 17.1
Which hormone is produced by the pancreatic beta cells in response to a rise in plasma glucose?

Question 17.2
Which hormone produced by pancreatic alpha cells stimulates the breakdown of glycogen into glucose?

Question 17.3
What region of the brain acts as the primary temperature regulatory centre?

Question 17.4
What mechanism in humans acts to reduce heat loss from the skin when the temperature falls?

Question 17.5
Which specialised tissue do infants use to generate heat?

18 The endocrine system

> **BASIC CONCEPTS:**
> The endocrine system acts to regulate a variety of functions in the body by the release of chemical messengers into the bloodstream. These messengers, or hormones, act on their target cells by binding to receptor molecules on the cell surface or in the cytoplasm. The binding of hormone to receptor triggers a cascade of cellular events which ultimately leads to a change in cell function. This can be achieved by enzyme activation, channel modification or changes in gene expression. Understanding the ways in which hormones exert their effects on target cells forms the basis of understanding many regulatory processes in the body.

18.1 Introduction

Hormones are traditionally defined as substances which are produced by cells within *endocrine glands* and which act on cells or tissues at a distance. Endocrine glands are collections of secretory cells which release their hormonal secretions into the bloodstream. Hormones interact with receptor molecules located either on the surface or within the cytoplasm of the target cell. They can produce immediate metabolic effects, as in the case of the stimulation of glucose uptake into liver and muscle by insulin (see Chapter 17). They can also have long-lasting effects on gene expression and cellular differentiation, as in the role of testosterone in the development of sexual characteristics. The *endocrine system* is defined as the set of organs or glands which produce hormones and includes the hypothalamus, pituitary, adrenal gland, thyroid gland, gonads, parathyroid glands and the pancreatic islets. Hormones are central to the body's homeostatic mechanisms and their production is often regulated via negative feedback loops (see Chapter 17).

It is now recognised that almost every cell in the body produces signal molecules which are released into the extracellular fluid, and which may influence other cells either nearby or at a distance. Many of these signal molecules are not described as hormones but may be known by other names such as cytokines, modulators or transmitters. An example of a non-classical hormonal system is the production of leptin by adipose tissue. Levels of this hormone are related to levels of stored fats and it acts on the brain to suppress appetite.

18.2 Types of hormone

There are three main classes of hormone – *steroid*, *polypeptide* and *amino acid* – but other molecules, such as glucose, may also be effective in cell signalling.

Steroid family hormones

Examples are testosterone, oestrogen and corticosterone. These are small molecules synthesised primarily from cholesterol which are released immediately following synthesis. They are hydrophobic and circulate in the blood bound to carrier proteins. They diffuse readily through cell membranes and act on intracellular receptors, which then bind to DNA and alter gene expression. In general they have slow, longlasting effects.

Polypeptide or protein hormones

These consist of chains containing between 3 and 332 amino acids. They are often synthesised as inactive precursors and stored prior to modification and release. They act on cell surface receptors then via second messenger systems (see Section 18.4). Examples are insulin and gastrin.

Amino acid hormones

Examples are thyroid hormone, noradrenaline and adrenaline. These are all derived from the amino acid tyrosine and are stored within their cells of production for immediate release. They have a variety of modes of action. Thyroid hormone acts through an intracellular receptor whereas the others act through cell surface receptors.

18.3 Cell signal receptors

The function of receptors is to act as the first stage in the process of converting the presence of the signal molecule into a cellular response. This process is known as *signal transduction*. There are two main classes of receptor – *cell surface and intracellular*.

Cell surface receptors

These are proteins which span the cell membrane and whose extracellular domain operates as a specific binding site for the signal molecule. The binding of the signal molecule to the binding site initiates a change in the receptor which alters the activity of the intracellular domain of the receptor. Changes to the intracellular domain then trigger an intracellular cascade of reactions which ultimately lead to the cellular response. Some receptors may be linked directly to channels and binding of the signal molecule will cause a change in the open–closed configuration of the channel, often leading to an influx of Ca^{2+} ions.

Intracellular receptors

These receptors are located in the cytoplasm or nucleus of the cell. Their active DNA-binding site is normally blocked by the presence of an inhibitory protein. They respond to lipid-soluble hormones which can pass directly through the cell membrane. Binding of the hormone causes a conformational change which releases the inhibitory protein and exposes the DNA-binding site. The receptor then acts as a transcription factor to initiate gene expression (see above and Chapter 5).

18.4 Second messenger systems

The binding of the signal molecule to a cell surface receptor may be a transient event lasting a few milliseconds. This may often lead to a cellular response which lasts minutes or hours. This process of signal amplification is achieved by the activation of second messenger systems. Second messengers are molecules whose levels are raised within the cytoplasm as a result of the interaction between the signal molecule and the receptor, and which, in turn, interact with intracellular proteins to produce a cellular response. A simple example of this principle can be seen in the regulation of smooth muscle contraction (see Chapter 16) where Ca^{2+} ions, which enter the cytoplasm either from outside the cell or from intracellular stores, interact with a calciumbinding protein which, in turn, interacts with the proteins of the contractile apparatus of the cell. An initial, extremely rapid depolarisation event can lead to a slow and sustained contraction.

G-protein linked receptors

G-proteins are small multi-unit proteins which are found on the cytoplasmic surface of the plasma membrane in close association with the intracellular domain of receptors. In the inactive state they bind guanosine diphosphate (GDP). Binding of the signal molecule to the receptor leads to a change in the conformation of the G-protein, which now exchanges GDP for GTP (guanosine triphosphate). This leads to detachment of the G-protein from the receptor. The G-protein then diffuses in the plane of the cell membrane until it binds to a target protein. This may result in either the modification of an ion channel or the activation of an enzyme. The activated enzyme then produces a second messenger molecule that enters the cytoplasm and interacts with its target proteins leading to a cellular response. Eventually the G-protein converts the bound GTP into GDP and is restored to its inactive state.

An example of this mechanism is seen in the kidney (see Chapter 24) where antidiuretic hormone (ADH) acts through a G-protein-linked receptor to activate the membrane enzyme adenylate cyclase. This enzyme converts ATP into a second messenger molecule, cyclic AMP (cAMP). This then acts through a cascade of protein intermediaries to cause the insertion of water channels into the cell membrane of the collecting duct – this promotes the re-uptake of water, preventing excessive fluid loss.

Tyrosine kinase receptors

A number of receptors, of which the insulin receptor is an example, have an intracellular domain which acts as a tyrosine kinase. A tyrosine kinase is an enzyme which adds phosphate groups to the amino acid tyrosine in other proteins. Phosphorylation alters the conformation of the recipient protein and, in most instances, causes its activation (see Chapter 2). The initial target proteins that are phosphorylated are often themselves kinases which, in turn, activate a further subset of proteins. This cascade effect leads to signal amplification, both in time (phosphorylated proteins may stay active for hours) and in amplitude (one receptor will activate many thousands of target proteins). In the case of the insulin receptor, phosphorylation of proteins in the cell may lead to the insertion of a glucose transporter into the cell membrane and to the activation of key proteins in the glucose→glycogen metabolic pathway.

18.5 Endocrine glands

The hypothalamus and the pituitary gland provide central control of a number of endocrine organs (Fig. 18.1). There is a hierarchy of systems in which the hypothalamus operates at the highest level and in which feedback to each level from the levels below acts to regulate hormonal production (see Fig. 18.3). The hypothalamus is a region of the brain which contains groups of neurons whose function is to provide homeostatic control over important bodily functions. For example, the hypothalamus regulates plasma osmolarity, appetite and sexual drive. In addition, the hypothalamus provides a link between higher consciousness and emotional responses through its regulation of the autonomic nervous system (see Chapter 19) and controls stress reactions through regulation of cortisol production by the adrenal gland.

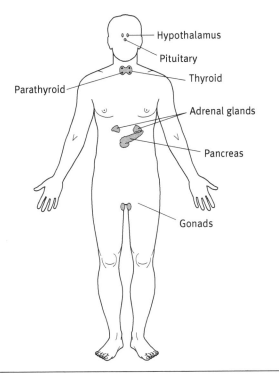

Figure 18.1. The endocrine glands

18.6   Hypothalamo-pituitary axis

Suspended beneath the hypothalamus and connected to it by the pituitary stalk is the *pituitary gland*. The pituitary gland is divided into anterior and posterior lobes. The anterior pituitary contains a collection of endocrine cells known as '...troph' cells. They are called this because the primary function of their secretions is to stimulate endocrine glands and other tissues elsewhere in the body. The '...troph' cells produce several protein hormones – *thyroid stimulating hormone, growth hormone, adrenocorticotrophic hormone, luteinising hormone, follicle stimulating hormone,* and *prolactin*.

Thyroid stimulating hormone (TSH)

Produced from thyrotrophs, TSH acts to stimulate the thyroid gland to produce the thyroid hormones T_3 and T_4. These are essential in normal development and promote increases in the basal metabolic rate of cells. Thyroid hormones act at nuclear receptors to alter gene expression. Lack of thyroid hormone will produce growth and mental retardation in the child and a sense of tiredness in the adult.

Growth hormone (GH)

Produced from somatotrophs, GH acts on multiple target tissues, including liver and muscle, to promote cell growth and enhance cellular metabolism. GH can act either directly or by stimulating the release of growth factors such as insulin-like growth factor-1.

CLINICAL EXAMPLE: ACROMEGALY

Excessive production of growth hormone following puberty, often as the result of a tumour of the anterior pituitary, results in excessive growth of bone and soft tissues. It is characterised by a thickening of the bones of the skull and by growth of skin. There is not normally an increase in height as the growth zones at the ends of long bones are sealed by this time. An individual with acromegaly will have a larger than normal head with coarse features.

Adrenocorticotrophic hormone (ACTH)

Produced from corticotrophs, ACTH acts on the adrenal gland to stimulate the production of the glucocorticoid hormone, cortisol. Cortisol has multiple effects including mobilisation of glucose from intracellular stores, suppression of the immune system, anti-inflammatory activity, reduction in bone production and effects on the central nervous system. Levels of cortisol are raised in chronic stress and analogues of cortisol are used as anti-inflammatory drugs.

Luteinising hormone (LH) and follicle stimulating hormone (FSH)

Produced from gonadotrophs, these hormones are the primary regulators of the ovary in females and the testis in males. In females, FSH promotes development of the ovarian follicle which surrounds the ovum and stimulates cells in the ovary to produce oestradiol. LH triggers ovulation and supports the formation of the corpus luteum, which is necessary for the development of the lining of the uterus in readiness for the implantation of the fertilised ovum. In the male, LH stimulates Leydig cells in the testis to produce testosterone, while FSH stimulates Sertoli cells to produce a number of factors which support sperm production.

Prolactin (PRL)

Produced by lactotrophs, PRL promotes the development of the breasts during pregnancy, initiates milk production and helps to maintain milk production once it has been established.

All of the '...troph' cells in the anterior pituitary are directly under the control of the hypothalamus. Hormones produced in the hypothalamus, and known as releasing hormones, are secreted into a special capillary network which connects the hypothalamus to the anterior pituitary. These hormones, such as gonadotrophin releasing hormone (GnRH) which acts on the gonadotroph cells, are often released in a pulsatile fashion and are subject to negative feedback from the hormones whose production they stimulate.

18.7 Posterior pituitary

The posterior pituitary is part of the neuroendocrine system. Neurons with their cell bodies in two nuclei in the hypothalamus send their axons down the pituitary stalk where they release hormones into capillaries in the posterior pituitary (Fig. 18.2). One of these nuclei produces the antidiuretic hormone (ADH) which acts on the collecting duct of the kidney to increase water reabsorption (see above). The other nucleus produces oxytocin which stimulates contraction of the uterus during the act of giving birth and milk release during breastfeeding.

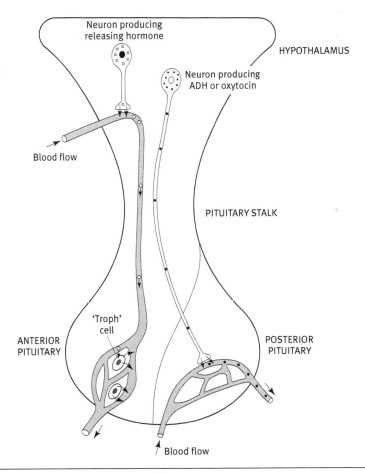

Figure 18.2. The pituitary and hypothalamus

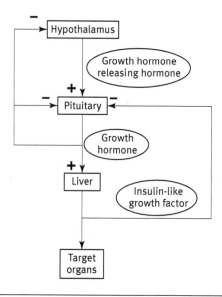

Figure 18.3. Growth hormone as an example to illustrate the hierarchy of systems and feedback in the endocrine system

18.8   Other endocrine glands

Parathyroid glands

There are four parathyroid glands located on the back of the thyroid gland. They produce parathyroid hormone (PTH) in response to falling plasma Ca^{2+} ion levels. The actions of PTH are to promote Ca^{2+} ion reabsorption in the kidney and to stimulate Ca^{2+} ion release from bone. This increases plasma calcium.

Endocrine pancreas

Within the pancreas are clusters of cells known as pancreatic islets. These contain α- and β-cells. The β-cells detect plasma glucose levels and respond to rising plasma glucose by secreting insulin. Insulin triggers the uptake and storage of glucose in liver and muscle cells. Glucagon, which is produced by the α-cells, acts to promote glucose production and release from the liver and is secreted at high levels following fasting.

Adrenal gland

In addition to cortisol the adrenal gland is responsible for the production of at least two other important hormones. The first is aldosterone which is involved in the regulation of extracellular fluid volume; it does this by promoting Na^+ ion, and thus water, reabsorption in the kidney (see Chapter 24). The second is adrenaline (epinephrine) which is produced in the centre of the adrenal gland by cells which are the equivalent of postganglionic neurons in the sympathetic nervous system. Adrenaline is released directly into the bloodstream and integrates the 'fight or flight' reaction (see Chapter 19) throughout the body.

18.9   Test yourself

The answers are given on pp. 176–177.

Question 18.1
Which family of hormones are small lipid molecules that act on intracellular receptors?

Question 18.2
Which hormone is responsible for the 'fight or flight' response?

Question 18.3
What nucleotide is bound to G-proteins when they are in the active state?

Question 18.4
From which hormone, produced by the adrenal glands in response to stress, are many anti-inflammatory drugs derived?

Question 18.5
List: (a) six hormones produced by the anterior pituitary gland; (b) two hormones released from the posterior pituitary gland.

19 The nervous system

BASIC CONCEPTS:

Nerve cells act to coordinate information received from multiple sources and to transmit this information to other cells. This ability of nerve cells to integrate information lies at the core of the functioning of the nervous system. Sets of nerve cells are organised into functional units which deal with specific types of information. Some may process sensory information, while others handle motor functions. Part of the nervous system is important in homeostasis. Understanding the basic structural patterns which underlie the organisation of the nervous system is important in appreciating its complexity.

19.1 Introduction

The nervous system is a complex assembly of nerve cells (see Chapter 16) and supporting glial cells which functions to control and coordinate all aspects of behaviour. The nervous system is responsible for higher thought processes, csoordination of movement, integration of sensory information, regulation of levels of consciousness, control of emotion and memory storage as well as control of essential bodily functions such as breathing.

19.2 Structure of nervous system

The nervous system comprises the *brain*, *spinal cord* and *peripheral nerves*. The brain is contained within the *cranial cavity* of the skull, and the spinal cord runs within the *vertebral column* (backbone) – see Fig. 19.1. Nerves run out from both the brain (*cranial nerves*) and the spinal cord (*spinal nerves*). *Motor neurons* convey information from the brain and spinal cord to muscles. *Sensory neurons* carry information from the sense organs to the spinal cord and brain.

An important functional classification is into the central nervous system (CNS) and the peripheral nervous system (PNS). The CNS includes all those neurons whose cell bodies, dendrites and axons lie entirely within the brain and spinal cord. The PNS includes neurons whose cell bodies and/or axons lie outside the brain and spinal cord. The differentiation into CNS and PNS is particularly important in terms of the ability to recover following damage, since regrowth of damaged axons can occur in the PNS but not within the

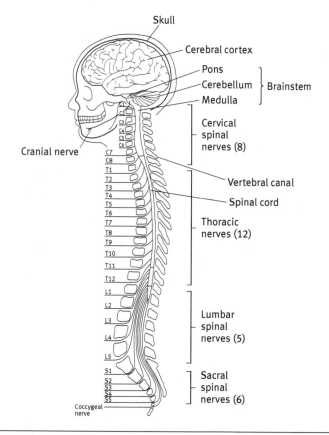

Figure 19.1. The brain and spinal cord

CNS. This has profound implications for patients who suffer damage to the nervous system.

Peripheral nerves are composed of the myelinated axons of motor and sensory neurons. Within the CNS, grey matter and white matter can be distinguished. *Grey matter* contains predominantly cell bodies and dendrites of neurons. Collections of nerve cells are known as *nuclei* or *ganglia* depending on their location. *White matter* contains myelinated axons running in tracts which connect one set of neurons to another.

19.3 The brain

The brain integrates information received from a variety of sources and produces a coordinated response. The brain can be divided into distinct components – the *cerebral cortex*, the *midbrain* and the *brainstem*.

Cerebral cortex

This is a folded sheet of neurons which covers the outer surface of the two hemispheres of the brain. The cerebral cortex is responsible for higher cognitive functions such as speech, thought and memory. The cerebral cortex is divided into four lobes (Fig. 19.2) each of which is specialised for particular functions.

- The *frontal lobe* is involved in planning movement and in regulation of emotions. On the left-hand side of the brain the frontal lobe contains a region responsible for speech production.
- The *parietal lobe* receives sensory information regarding pain, touch and temperature from the entire body and also acts to integrate this with other sensory information such as vision, hearing, taste and smell.
- The *occipital lobe* at the back of the brain is the area into which visual information is received and where processing of this information takes place.
- The *temporal lobe* contains areas associated with emotion and the regulation of emotional responses. It enables the formation of memory. The temporal lobe also receives hearing and smell sensory input.

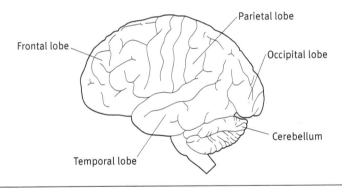

Figure 19.2. The cerebral cortex

The two hemispheres of the brain are connected by a white matter bridge known as the *corpus callosum* which ensures integration between the two sides. In general, the right-hand side of the brain receives sensory input from and sends motor output to the left side of the body, and the left-hand side of the brain controls the right side of the body.

Midbrain

This region contains important sets of nuclei which act to link the cerebral cortex with the brainstem and spinal cord. The *thalamus* contains a number of nuclei which relay sensory information towards the cortex. The *basal ganglia* are involved in the planning of movement. Also in the midbrain is the *hypothalamus* which contains nuclei involved in homeostasis and endocrine functions (see Chapters 17 and 18).

Brainstem

This comprises the *pons*, *medulla* and *cerebellum*. The pons and medulla contain important nuclei associated with sensory and motor functions of the cranial nerves (such as movement of the eyes and tongue, taste, control of facial expression and chewing, and touch, pain and temperature sensation from the face) as well as key nuclei involved in respiratory and cardiovascular regulation. The cerebellum ensures that movements are carried out smoothly and helps to regulate balance.

19.4   The spinal cord

The primary function of the spinal cord is to convey information from the brain to the rest of the body, and from the rest of the body to the brain. The spinal cord gives rise to 31 pairs of spinal nerves (Fig. 19.1). Each spinal nerve is formed from the fusion of dorsal and ventral roots which arise from the spinal cord (Fig. 19.3). The dorsal roots carry sensory axons into the spinal cord. The ventral roots convey motor axons away from the spinal cord. Motor axons originate from motor neurons whose cell bodies lie within the ventral horn of the central grey matter of the spinal cord. The motor neurons are controlled by input from the cerebral cortex and from brainstem nuclei, and their axons will eventually terminate at neuromuscular junctions on skeletal muscle.

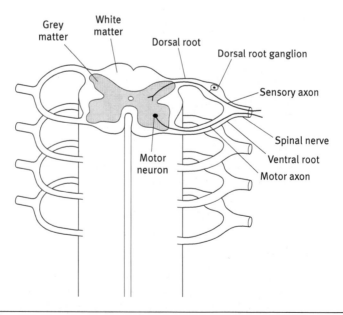

Figure 19.3. The spinal cord – front view

19.5 Peripheral nervous system

The peripheral nervous system is divided into *somatic* and *autonomic* divisions. The somatic division contains sensory fibres from the skin and special sense organs such as the eye and the ear. It supplies motor innervation to skeletal muscle. The autonomic nervous system is involved in the control of glands, the heart and visceral smooth muscle.

Somatic division

Many spinal nerves, once they have emerged from the vertebral column, join up with spinal nerves from higher and lower levels of the spinal cord to form nerve *plexuses*. In these plexuses, axons from various spinal nerves join together to form a single peripheral nerve. This nerve will then run through the body to reach its target regions. As an example, the *sciatic nerve* is formed from the fusion of five spinal nerves which emerge from the lower regions of the vertebral column. The sciatic nerve runs down the back of the leg and supplies the muscles of the back of the leg above the knee and all the muscles of the leg below the knee. Sensory nerve endings in the skin and in muscles and joints of the leg give rise to axons which run back in the sciatic nerve to enter the spinal cord through the dorsal roots of the individual spinal nerves which comprise it. The cell bodies of these sensory axons are clustered in special ganglia associated with the dorsal roots.

Autonomic division

The autonomic nervous system is divided into *sympathetic* and *parasympathetic* divisions. Both divisions tend to innervate smooth muscle and glandular tissue and in many instances have opposite effects on their target tissues (Fig. 19.4).

Sympathetic nerves contain axons originating from a series of ganglia which lie alongside the vertebral column. These ganglia are controlled by axons which arise from neurons in the thoracic or lumbar regions of the spinal cord. Sympathetic nerves use *noradrenaline* as their neurotransmitter at the target organ. They produce physiological changes which characterise the 'fight or flight' reaction. These include:

- increased heart rate and force of contraction of the heart
- dilation of the arteries to the heart
- decreased blood flow to the skin
- increased blood flow to muscles
- dilation of the airways
- sweating
- piloerection (hair stands on end)
- reduced blood flow to the GI tract

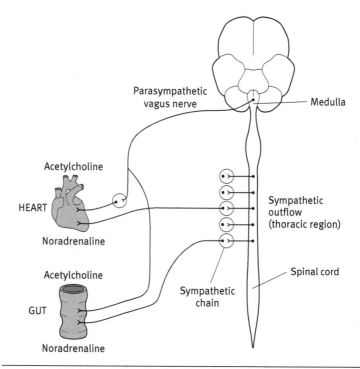

Figure 19.4. Innervation of the heart and GI tract by the autonomic nervous system

- production of a viscous saliva
- stimulation of the adrenal gland to release adrenaline into the bloodstream.

Parasympathetic nerves contain axons which originate from the brainstem or the sacral region of the spinal cord. They synapse with neurons in ganglia very close to the target organs and the axons of these ganglionic neurons use acetylcholine as their neurotransmitter at the target organ. The vagus nerve, which originates in the brainstem, is the main parasympathetic nerve in the body and supplies all thoracic and abdominal organs.

The physiological changes produced by the parasympathetic system can be characterised as 'rest and digest' and include:

- decreased heart rate
- increased blood flow to the GI tract
- increased acid production in the stomach
- increased production of a watery enzyme-rich saliva
- constriction of the airways.

CLINICAL EXAMPLE: STROKE

Blockage of arteries bringing blood to the brain causes the death of neurons in the area of the brain supplied. This is known as a stroke and leads to a loss of function in the parts of the body controlled by the affected cells. A common site for stroke is the area supplied by the middle cerebral artery which includes large sections of the frontal and parietal lobes of the cerebral cortex. This leads to loss of motor and sensory functions on the opposite side of the body and, if the lesion is in the left hemisphere of the brain, to a loss of the ability to speak.

19.6 Test yourself

The answers are given on p. 177.

Question 19.1
What type of neurons carry information from the central nervous system to muscles?

Question 19.2
What names are given to the four lobes of the cerebral cortex?

Question 19.3
Which division of the peripheral nervous system innervates skeletal muscle and the sense organs?

Question 19.4
Which neurotransmitter is used by: (a) sympathetic nerves; (b) parasympathetic nerves at their target organs?

Question 19.5
What is the main parasympathetic nerve supplying the organs of the thorax and abdomen?

20 The cardiovascular system

> **BASIC CONCEPTS:**
> Circulation of the blood around the body is vital to ensure the supply of oxygen and nutrients to the tissues and the removal of waste. Oxygen is carried by haemoglobin within red blood cells. The white blood cells are key components in the body's defence mechanisms against infection. Circulation of blood through the arteries and veins is accomplished by the pumping action of the heart. The activity of the heart is closely regulated to enable sufficient pressure to be maintained in the circulation to ensure an adequate blood supply to all tissues, especially the brain. An understanding of the factors which are involved in the regulation of the cardiovascular system is essential to an appreciation of its central role in the maintenance of life.

20.1 Introduction

The cardiovascular system is designed to ensure the circulation of oxygenated blood to all organs of the body. Oxygenated blood from the lungs (see Chapter 21) is returned to the left side of the heart in the pulmonary veins and is then pumped out through the aorta into the systemic circulation in which it is distributed to the organs of the body via blood vessels known as *arteries*. Blood flows through these organs in narrow diameter, highly permeable capillaries. These facilitate exchange of substances between blood and the interstitial fluid bathing the cells of the organ. Deoxygenated blood is then returned, in blood vessels known as *veins*, to the right side of the heart. The blood is then pumped back to the lungs via the pulmonary arteries. Blood in the systemic circulation conveys nutrients and oxygen to the tissues of the body and carries away waste products. Blood also acts to transport chemical signalling molecules (hormones) from their site of production (endocrine glands) to their sites of action.

20.2 Blood

Blood has three main constituents:

- plasma (50–60% of volume)
- red blood cells (40–50% of volume)
- white blood cells and platelets (1% of volume).

The white blood cells are part of the immune system and are considered in more detail in Chapter 25. Blood is circulated around the body within blood vessels and functions primarily to transport essential substances from one organ to another.

20.3 Plasma

Plasma comprises a protein-rich solution in which the cellular constituents of the blood are suspended. The ionic composition of plasma is similar to that of interstitial fluid and it has an osmolarity of 290 mosmol/l, of which Na^+ and Cl^- are the major ions. Plasma also contains K^+ ions at 4 mmol and Ca^{2+} ions. Much of the plasma calcium exists not as free ions but in a bound form attached to the major plasma protein albumin. Minor fluctuations of plasma potassium can have significant consequences for excitable tissues as the resting membrane potential is predominantly determined by the K^+ ion concentration gradient across the cell membrane (see Chapter 16) .

Apart from the immunoglobulins the majority of plasma proteins are synthesised in the liver. Plasma proteins have a variety of functions which are now outlined.

Transport

Many substances are transported in blood bound to plasma proteins. Some, such as steroid hormones, are transported attached to albumin, the most common plasma protein. Others have their own specific carrier proteins, such as iron which is transported bound to ferritin.

Defence against infection

See Chapter 25 for a full discussion of this function.

Blood clotting

It is important that blood vessels, if they become damaged, do not continuously leak and allow significant blood loss. Plasma contains a group of proteins and some microcellular components known as *platelets* which work cooperatively to produce a blood clot to plug damaged vessels. In general, clotting is initiated by exposure of specific proteins in the epithelial lining (endothelium) of damaged vessels. The final component of the protein-clotting cascade is a filamentous protein, *fibrin*, which is cross-linked by the action of the enzyme thrombin. The fibrin meshwork that is formed traps platelets which help to build up the plug. Platelets, in turn, release the signal molecule serotonin which acts to cause neighbouring blood vessels to constrict and further minimise blood loss.

> **CLINICAL EXAMPLE: HAEMOPHILIA**
>
> One of the proteins of the clotting cascade is Factor VIII. This is a vital component in the chain reaction leading from endothelial damage to blood clot formation. The gene coding for Factor VIII is present on the X chromosome and mothers with one copy of the faulty gene have a 50% chance of passing this to their sons. Since boys have only one X chromosome those that do inherit the faulty gene will not synthesise any Factor VIII and will not be able to form blood clots. Even small wounds are potentially fatal to sufferers of this condition. Haemophilia has been treated by infusion of Factor VIII purified from the blood of donors – it has led to problems when the donated blood also contained infections such as HIV or CJD.

Oncotic pressure

The presence in plasma of proteins, such as albumin, which carry a high negative charge helps to create what is known as plasma oncotic pressure. This is a combination of the osmotic forces due to the protein particles themselves, which cannot pass between capillary and interstitial fluid, and additional osmotic forces due to the cations which these proteins attract and hold. Plasma oncotic pressure is equivalent to 25 mmHg and is the key force drawing fluid back into the venous end of capillaries once it has circulated through the interstitial spaces. In situations where plasma oncotic pressure is reduced, such as in liver failure when plasma protein levels drop, then fluid stays in the interstitial space resulting in tissue swelling (known as *oedema*).

20.4 Red blood cells

Red blood cells (*erythrocytes*) are continuously produced in the bone marrow. Their formation is under the control of the hormone erythropoietin which is produced by the kidney in response to decreasing arterial oxygen levels. They are shaped as a biconcave disc which allows both a large surface area to volume ratio and flexibility as they pass through small capillaries. Red blood cells do not have a nucleus and the cytoplasm is packed with the protein *haemoglobin*, which serves to transport oxygen around the body. Erythrocytes stay in circulation for about 150 days before being broken down in the spleen.

Haemoglobin is a protein consisting of four subunits (α-globin chains) each of which has at its core a haem ring. This is a structure containing reduced iron which allows the binding of oxygen. There is an interaction between the four subunits such that when O_2 is bound to one unit it becomes easier for O_2 to bind to the remaining subunits. The binding of O_2 to haemoglobin is normally represented graphically (Fig. 20.1) by plotting the partial pressure of oxygen (pO_2) against the % saturation (amount bound/total capacity $\times$ 100) of the haemoglobin. This sigmoidally-shaped graph shows that at a normal lung pO_2 of 100 mmHg the haemoglobin will be 97.5% saturated, whereas at a normal tissue pO_2 of 40 mmHg the saturation will drop to 70%. Thus when haemoglobin within erythrocytes arrives in tissues, approximately 25% of the carried oxygen will dissociate from the haemoglobin and diffuse into cells.

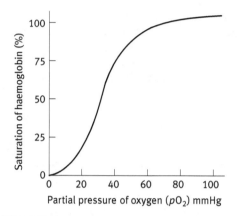

Figure 20.1. Haemoglobin dissociation curve

20.5 Platelets

Platelets are small subcellular fragments formed by being pinched off from their precursor cell, the *megakaryocyte*. They are important in the formation of blood clots (see Section 20.3).

20.6 The heart

Structure

The heart is a four-chambered pump which is responsible for maintaining blood flow within the systemic and pulmonary circulations – Fig. 20.2 shows a schematic diagram of its structure. The four chambers comprise two *ventricles* and two *atria* (singular *atrium*). The activity of the heart is divided into two phases – *systole*, during which the ventricles contract, and *diastole*, during which the ventricles relax and the atria contract.

The right atrium of the heart receives venous blood back from the systemic circulation via veins called the *superior and inferior venae cavae*. The right atrium is separated from the right ventricle by a valve (the *tricuspid valve*) which opens when the pressure in the ventricle is lower than that in the atrium. This occurs at the end of systole as the ventricle relaxes, and filling of the ventricle is achieved by a combination of passive flow of blood from the

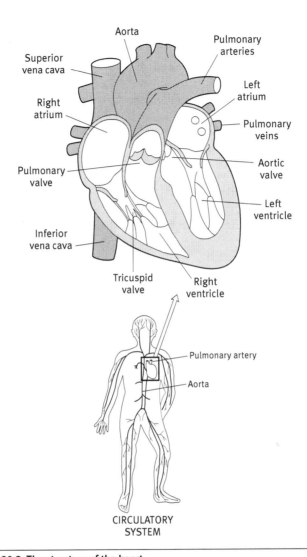

Figure 20.2. The structure of the heart

atrium and weak contraction of the atrial wall. Once ventricular filling is complete, the ventricle will begin to contract and blood is pushed out through a valve at the base of the *pulmonary trunk* into the *pulmonary arteries* and into the lungs. The increased blood pressure in the ventricle causes closure of the tricuspid valve which prevents backflow of blood into the atrium.

Venous blood return from the lungs is via the *pulmonary veins* which empty into the left atrium. This is connected to the left ventricle through the *bicuspid* or *mitral valve* and left ventricular filling also occurs during diastole. During systole left ventricular contraction forces blood into the aorta through the aortic valve at its base. The volume of blood ejected is known as the *stroke volume* and is approximately 50 ml.

The aorta and pulmonary trunk have elastic tissue in their walls and expand to accommodate the blood ejected from the heart. During diastole there is elastic recoil of the walls of these large vessels and this ensures the continued circulation of blood around the body or to the lungs while the ventricles are relaxing.

Function

The timing of the events of the cardiac cycle is achieved via specialised excitable tissue within the wall of the heart (see Chapter 16). The structure of cardiac muscle is such that the gap junctions between adjacent fibres provide electrical continuity, which means that depolarisation and contraction of any one cardiac muscle fibre will spread to all adjacent fibres. The muscle fibres of the atrial and ventricular walls are separated by a non-conducting fibrous *septum* but within either atria or ventricles a single point of depolarisation will produce contraction of all the muscle fibres.

The *sinu-atrial* (S-A) *node*, which is located in the wall separating the right from the left atrium, contains cells which are capable of generating spontaneous action potentials at a regular rate of about 70/min. The cells of the S-A node are connected via gap junctions to the muscle fibres of the atrial walls and so are able to initiate synchronised atrial contractions. The wave of depolarisation which spreads through the atrial wall is picked up by another specialised set of cells, the *atrio-ventricular* (A-V) *node*. This node transmits the action potential to the ventricles via a set of elongated muscle fibres known as the *A-V bundle*. This pierces the fibrous septum separating the atria from the ventricles and travels down the dividing wall between the two ventricles until it reaches the point of the ventricles furthest away from the aorta and pulmonary trunk. Contact is then made with the muscle fibres of the ventricular wall and contraction is initiated. There is a delay between the wave of excitation arriving at the A-V node and the initiation of ventricular contraction that allows time for filling of the ventricles to be completed.

The electrical events of the heart can be recorded by placing electrodes on the surface of the chest. The resultant trace is the electrocardiograph (ECG). An ECG (Fig. 20.3) has three main elements:

- a P wave which represents atrial depolarisation;
- this is followed after a delay by the QRS complex which represents ventricular depolarisation;
- finally a T wave which indicates ventricular repolarisation.

ECG trace showing rhythmic beating of heart

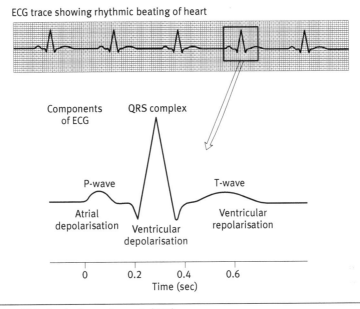

Figure 20.3. An electrocardiograph (ECG)

20.7 The circulatory system

The circulatory system consists primarily of *arteries* which take blood away from the heart and *veins* which return the blood to the heart. The two sides of the circulation are connected by *capillaries* which permit exchange between blood and tissues. Blood leaving the heart enters the *aorta* which contains many layers of elastic tissue in its wall. From here the blood passes to muscular arteries, such as the radial artery in the arm (from which the pulse rate is normally measured at the wrist). These arteries control the distribution of blood to the various organs of the body. Muscular arteries branch repeatedly to lead to narrow *arterioles* which connect to the capillary beds. The tone of the smooth muscle in the walls of the arteries and arterioles is controlled through the sympathetic nervous system.

Because of their narrow diameter and large total cross-sectional area, arterioles are the site of the major peripheral resistance to blood flow. From the capillaries blood passes back into the thin walled *venules* and *veins* for return to the heart.

The heart must work to force blood through the circulatory system against the resistance offered by narrow peripheral vessels. This generates a pressure in the circulatory system known as *blood pressure*. Blood pressure is measured during both systole and diastole. It is normally expressed as two figures – for example 120/80 mmHg. The first figure is the systolic pressure and is an indication of the strength of ventricular contraction. The second figure is the diastolic pressure and relates to peripheral resistance. The factors which influence blood pressure are described by the formula:

$$\text{blood pressure} \propto \text{heart rate} \times \text{stroke volume} \times \text{peripheral resistance}.$$

This means that if the amount of blood pumped by the heart or the resistance to its flow increase then blood pressure will rise.

Regulation of blood pressure can be achieved by three key mechanisms. One is a *neurogenic* mechanism which involves *baroreceptors* (pressure receptors) located in the aorta and arteries in the neck. Decreasing blood pressure leads to an increase in signals sent to the cardiovascular centre in the medulla of the brain. This in turn triggers an increase in sympathetic output resulting in an increase in heart rate and an increase in peripheral resistance as the result of vasoconstriction. Increasing blood pressure will reverse these effects. The second mechanism involves the kidneys where a decrease in blood pressure in arterioles leads to the release of the enzyme renin. This converts the circulating hormone precursor angiotensinogen to angiotensin I. Angiotensin I is then converted by an enzyme in the walls of blood vessels to angiotensin II, which acts to increase peripheral vasoconstriction and thus raises blood pressure. A third regulatory mechanism involves regulation of fluid volumes by the kidney and will be dealt with in Chapter 24.

CLINICAL EXAMPLE: HYPERTENSION

A diastolic blood pressure consistently over 80 mmHg is classed as *hypertension*. A prolonged increase in blood pressure is linked with an increased risk of heart attack and stroke, as well as damage to the kidneys and to the eyes. The causes of hypertension are frequently unknown but risk factors may include obesity and stress. Lifestyle modifications, such as healthy eating and exercise, are commonly the first line of treatment. Pharmacological treatments may involve beta blockers to slow down the heart rate or inhibitors of the enzyme which converts angiotensin I to II (ACE inhibitors).

20.8 Test yourself

The answers are given on p. 177.

Question 20.1
What filamentous protein is cross-linked by thrombin to form clots?

Question 20.2
What name is used to describe the phases of the cardiac cycle in which: (a) contraction of the ventricles and (b) contraction of the atria occurs?

Question 20.3
What valve separates: (a) the right atrium from the right ventricle; (b) the left atrium from the left ventricle?

Question 20.4
Which chambers of the heart receive: (a) oxygenated blood returning from the lungs; (b) deoxygenated blood returning from the tissues?

Question 20.5
Which three physiological factors affect blood pressure?

21 The respiratory system

> **BASIC CONCEPTS:**
> The respiratory system ensures a constant supply of oxygen to the body and the removal of carbon dioxide. During respiration the lungs draw in air from outside the body and mix it with air held within small thin-walled sac-like structures known as alveoli. These are in close contact with pulmonary capillaries and this permits the rapid diffusion of oxygen from air to blood and of carbon dioxide in the reverse direction. Air is then returned to the atmosphere during expiration and waste carbon dioxide is thus removed. It is important to understand the factors which control the movement of air into and out of the lungs and the mechanisms of gaseous exchange.

21.1 Introduction

Cellular respiration (see Chapter 8) involves the utilisation of oxygen and the generation of carbon dioxide during the aerobic production of energy. Oxygen is carried to the tissues bound to the haemoglobin in red blood cells (see Chapter 20) and carbon dioxide is transported away in a variety of forms but mainly as bicarbonate ions formed as a result of the reversible reactions

$$H_2O + CO_2 \leftrightarrow H_2CO_3 \leftrightarrow H^+ + HCO_3^-$$

The respiratory system is designed to ensure that the blood remains oxygenated and that excess CO_2 is removed to the atmosphere.

21.2 Structure of the respiratory system

The respiratory organs in mammals are the *lungs*, which lie within the *thoracic cavity*. The surface of each lung is covered by a thin *pleural membrane* which is continuous with the pleural membrane lining the inside of the wall of the thoracic cavity. This arrangement is best visualised by imagining a partially inflated balloon (the pleura) contained within a box (the thorax). If you were to push your fist (representing the lung) into the balloon then there would be a layer of balloon coating your fist and another layer lining the inside of the box, and between them would be an air-filled space, the pleural cavity (Fig. 21.1). In reality this pleural cavity, which lies between the lung and the wall of the thorax, contains only a small amount of fluid and the two layers of pleural membrane are closely related to each other.

Because the pleural cavity is a closed space any movement of the thoracic wall is coupled to movement of the outer surface of the lung within it. Thus when the volume of the thorax is increased during inspiration (see below) the volume of the lung is increased by an equivalent amount and air is drawn in.

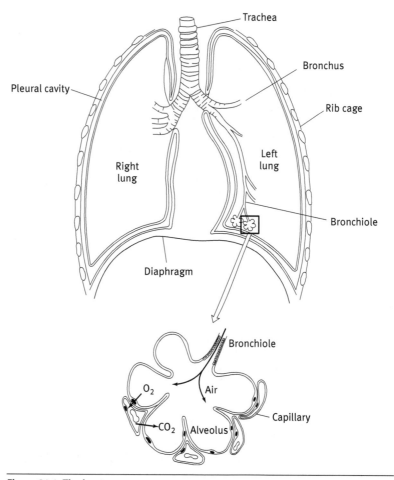

Figure 21.1. The lungs

The respiratory system is divided into two main parts – the *conducting portion* and the *respiratory portion*.

Conducting portion

The function of the conducting portion of the respiratory system is to moisten, warm and clean inspired air before it is delivered to the respiratory portion of the system where gaseous exchange between air and blood takes place.

Air is initially inspired through the nose and then passes through the *nasopharynx* at the back of the mouth, through the *larynx* (voicebox) and into the *trachea* (windpipe). From the trachea the airways branch repeatedly through *bronchi* and *bronchioles*, which become progressively narrower until they terminate at the small sac-like *alveoli* (singular *alveolus*) comprising the respiratory portion of the system. The wider parts of the airways are maintained open by rings of cartilage in their wall. The major resistance to airflow into the lungs is found in medium-sized airways and increasing the diameter of these airways by relaxation of smooth muscle in their walls is important when increased air flow into the lungs is required. The airways are lined by a specialised *respiratory epithelium* which contains a mixture of mucus-secreting goblet cells and *columnar cells* with hair-like protrusions called *cilia* (singular *cilium*). The mucus acts as a sticky layer to trap inhaled dust particles and bacteria. The cilia constantly beat to move the mucous layer up the airways towards the back of the throat where it is swallowed.

Respiratory portion

This is the part of the lung in which exchange of gases takes place between the inspired air and the blood entering the lung in the pulmonary artery. The basic structural unit is the alveolus which is a sac-like structure with walls made up of squamous cells (see Chapter 14). These cells share a basal lamina with the squamous endothelial cells of the pulmonary capillaries and together they form the diffusion barrier. This is normally less than 1.5 μm thick and permits rapid diffusion of both O_2 and CO_2 between blood and the air in the alveolus. There are approximately 300 million alveoli in the lungs with a total surface area of 80 m^2.

21.3 Respiration

With respect to the respiratory system, the term *respiration* is used to describe the act of breathing, through which air is moved into (*inspiration*) and out of (*expiration*) the lungs. Because the lungs are tightly linked to the walls of the thoracic cavity the volume of the lungs is increased by increasing the volume of the thoracic cavity. There are two ways in which this happens. The first involves movements of the *diaphragm*. This is a domed muscular sheet which separates the thorax from the abdomen and which, at rest, lies level with the lowest part of the *sternum* (breastbone). During inspiration the diaphragm moves downwards and so the height of the thoracic cavity is increased. The second way involves movements of the rib cage which act to increase the diameter of the thoracic cavity (Fig. 21.2). If, as you breathe in, you place your hand on the lowest part of your sternum you will feel it move upwards and outwards.

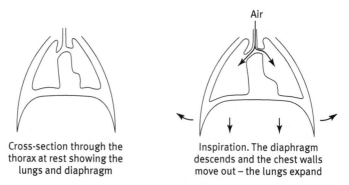

Cross-section through the thorax at rest showing the lungs and diaphragm

Inspiration. The diaphragm descends and the chest walls move out – the lungs expand

Figure 21.2. Movement of the diaphragm in respiration

Air in the lungs is normally equilibrated with atmospheric pressure but as the lungs expand then their internal pressure drops and air is drawn in. The lungs have an inherent resistance to expansion which is contributed to by having to stretch elastic fibres in the walls of the alveoli and by the surface tension of the squamous alveolar cells. Thus inspiration requires active muscular effort in the rib cage and the diaphragm to overcome this resistance. The ease with which a lung can be expanded is known as its *compliance*. Surface tension in the lungs is lowered, and thus compliance increased, by the secretion onto the surface of the alveolar cells of a lipid substance known as *surfactant*. Once the inspiratory muscles relax, the lungs recoil and return to their resting state. During this recoil phase air is forced out of the lungs.

21.4 Gaseous exchange

The levels of O_2 and CO_2 in blood and air are described in terms of their partial pressures. The SI unit of pressure is the pascal (Pa) and partial pressures can be expressed in kilopascals (kPa). However, because blood pressure is always expressed clinically in millimetres of mercury (mmHg) other physiological pressure measurements also tend to be expressed in this older unit. The partial pressure of oxygen in atmospheric air is written as pO2 and normally has a value of 160 mmHg. In oxygenated blood the partial pressure is written p_aO_2 and has a range 75–100 mmHg. Blood heading for the lungs in the pulmonary arteries has a p_vO_2 of around 40 mmHg and a p_vCO_2 of around 45 mmHg. During inspiration a fixed volume of air is drawn into the lungs (the tidal volume) and mixes with the amount of air left in the lungs at the end of expiration (the *residual volume*). This residual air has been partially depleted of oxygen and contains higher levels of carbon dioxide than are found in the atmosphere. As a result of the mixing of fresh inspired air with this air left over from previous breaths, the p_AO_2 in the alveoli at the end of inspiration is approximately 100 mmHg and the p_ACO_2 is approximately 40 mmHg. Diffusion of a gas between blood and air will occur when the pressures are unequal on either side of the diffusion barrier. Thus, under normal circumstances there will be diffusion of oxygen from the alveolar air into the capillary and of carbon dioxide in the reverse direction (Fig. 21.3). The amount of gas exchange that can occur between the capillary and the alveolar air is normally limited only by the rate of flow of blood through the capillaries – the greater the rate of flow, the more gas that can be exchanged.

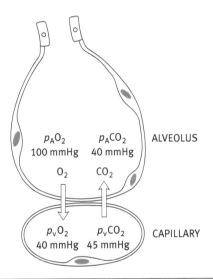

ALVEOLUS

p_AO_2 100 mmHg

p_ACO_2 40 mmHg

O_2 CO_2

CAPILLARY

p_vO_2 40 mmHg

p_vCO_2 45 mmHg

Figure 21.3. Gaseous exchange between air and blood in the lung

21.5 Control of respiration

Respiration is controlled by a group of neurons in the medulla of the brain known as the *respiratory centre* – this acts to control the activity of the muscles of respiration. The diameter of the airways is regulated via the autonomic nervous system. The respiratory centre neurons are directly activated by a rise in p_aCO_2 and this is the primary stimulus to increased respiratory effort. There are also sensors in the blood vessels of the head and neck which can detect low levels of oxygen.

> **CLINICAL EXAMPLE: ASTHMA**
>
> In asthma an external stimulus, such as pollen or animal hair, triggers an inflammatory reaction in the walls of the airways which causes them to constrict. This reduces the amount of air which can move in and out of the lungs and leads rapidly to an increase in p_aCO_2 since this cannot be removed from the system. Respiratory activity is increased and the affected individual will be seen to be gasping for air. However, because of the narrowed airways it is virtually impossible to increase the flow rate to the alveoli and eventually p_aO_2 levels will drop. In severe attacks the patient may lose consciousness. Treatment is by administration of inhaled drugs which relax the muscle of the airways and allow air to be drawn in.

21.6 Test yourself

The answers are given on p. 177.

Question 21.1
What name is given to the membrane that surrounds the lungs and lines the thoracic cavity?

Question 21.2
What are the two main types of cell in respiratory epithelium?

Question 21.3
What muscular sheet separates the thorax from the abdomen?

Question 21.4
Which cells make up the diffusion barrier between blood and air?

Question 21.5
To changes in what are cells in the medullary respiratory centre particularly sensitive?

22 The digestive system

> **BASIC CONCEPTS:**
> All energy is derived from the food we eat and in order for this energy to be used by the body these foodstuffs must be broken down into molecules that can be absorbed from the digestive tract into the bloodstream. The primary functions of the digestive tract are thus digestion and absorption. The digestive system consists of a hollow tube, the gastrointestinal tract, which extends from the mouth to the anus and which receives the secretions of various glands to aid the digestive processes. Some regions are specialised for digestion and others for absorption. Understanding the function of the digestive system is an important step in understanding how the energy requirements of the organism are met.

22.1 Introduction

Survival depends on adequate intake of food and water. Food provides the raw materials for energy production (see Chapter 8) as well as the building blocks for cells. In addition, food contains nutrients such as essential vitamins and minerals – for example vitamin B1 (see Chapter 8), iron for the production of haemoglobin (see Chapter 20) and calcium for the formation of bones (see Chapter 26). The functions of the digestive system can be summarised as *digestion*, *absorption* and *excretion*. Food is broken down into molecules small enough to be absorbed through the wall of the gut into the bloodstream. Unwanted or undigested components of the diet are then excreted along with other waste products.

The digestive system comprises the *gastrointestinal* (GI) *tract* which runs from the *oral cavity* (mouth) to the *anus* and includes the oesophagus, *stomach*, *small* and *large intestines* and *rectum*. Associated with the GI tract are a number of accessory glands (*salivary glands*, *pancreas* and *liver*) whose secretions enter the GI tract and play a vital role in the digestive process (Fig. 22.1).

22.2 Oral cavity

The process of digestion is initiated in the oral cavity where food is broken down into small pieces by the actions of the teeth and tongue during chewing. In the oral cavity, food is mixed with the secretions of salivary

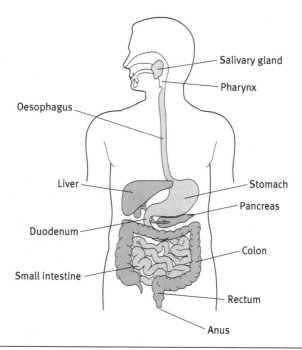

Figure 22.1. The gastrointestinal tract

glands, which produce a watery secretion containing the enzyme salivary α-amylase, which initiates the digestion of starch. The mixture of food and saliva forms a paste which is then passed into the oesophagus during the act of swallowing. The initial phase of swallowing is voluntary and depends on the actions of the tongue and of the muscles of the upper throat (*pharynx*).

22.3   Structure of the GI tube

The oesophagus connects the mouth to the stomach and is a hollow, muscular tube which forms the first part of the GI tract. The structure of the wall of the GI tube is relatively constant throughout its length and comprises three main layers which enclose the central space or *lumen* (Fig. 22.2). The innermost layer is the *mucosa* which lines the lumen with an epithelium. The structure of the epithelium varies depending on the functional demands of the various regions of the GI tract, but for the most part it is simple columnar. The mucosa is bounded internally by a layer of smooth muscle which allows the mucosa to be mobile and promotes local mixing of the contents of the lumen. Outside the mucosa lies a layer of connective tissue known as the *submucosa*, which may contain additional glands and which also contains collections of cells from the immune system which make up the gut-associated lymphoid tissue (GALT). The outermost part of the gut wall is the

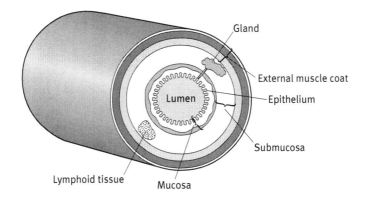

Figure 22.2. The gut tube

external muscle coat which contains two layers of smooth muscle – one running longitudinally along the length of the gut and the other arranged in a circular or spiral fashion. This muscle coat is responsible for the wave-like movements of the gut wall (*peristalsis*) which help to push food along the length of the GI tract. The movements of the external muscle coat are regulated by the GI tract's own nervous system (the *enteric* nervous system). This contains up to 10 million neurons (as many as are found in the spinal cord!) which act locally in a reflex fashion to ensure coordinated movement of the gut wall. The activity of the enteric nervous system can be modified by the autonomic nervous system (see Chapter 19) and by locally produced peptide hormones.

22.4 The stomach

Contractions of the oesophagus cause the swallowed food to enter the stomach. This is a sac-like expansion of the GI tube which lies just below the diaphragm in the upper-left quadrant of the abdominal cavity. The stomach is stretchable to allow it to accommodate the food taken in during a meal. The external muscle coat of the stomach is particularly thick and vigorous contractions produce a churning movement. The entrance and the exit of the stomach are controlled by thickened circular bands of muscle. The upper of these is called the *oesophageal sphincter* and prevents the stomach contents being forced back up into the oesophagus. Failure of this sphincter can allow acid gastric juices to enter the oesophagus where they cause a painful burning sensation – heartburn. The *pyloric sphincter* prevents the passage of food from the stomach into the *duodenum* (the first part of the small intestine) before stomach functions are complete. On average food spends 4–6 hours in the stomach. During this time it is mixed with the secretions of the gastric glands. These are extensions of the epithelium lining the stomach wall and contain two key cell types – *parietal* and *chief* cells.

Parietal cells

These are responsible for the secretion of gastric acid. This is hydrochloric acid whose secretion maintains a low pH in the stomach. Acid secretion by parietal cells is regulated by three factors:

- the sight and smell of food triggers the vagus nerve to release acetylcholine (ACh) in the stomach wall where it acts on receptors on parietal cells;
- the presence of protein in the stomach triggers the release of a hormone, gastrin, which also acts directly on parietal cells to increase acid secretion;
- both gastrin and ACh trigger the local release of histamine which acts on parietal cells to enhance acid release.

In addition, parietal cells secrete intrinsic factor which is required for the intestinal absorption of vitamin B_{12}, an essential compound in haemopoiesis.

Chief cells

These secrete the digestive enzymes pepsinogen, amylase, lipase and prorennin. All of these enzymes are active at acidic pH. Both pepsinogen and prorennin are converted to their active forms, pepsin and rennin, by the stomach acid and all work optimally at low pH. Pepsin initiates the digestion of proteins and rennin is particularly important in young animals where it breaks down milk proteins.

CLINICAL EXAMPLE: GASTRIC ULCERS

Gastric ulcers occur when the mucosal barrier in the stomach is overwhelmed by the amount of acid produced. Once the mucus is breached then the acid begins to erode the wall of the stomach. This initially causes discomfort, but if the erosion penetrates the wall of the stomach then nearby blood vessels can also be destroyed leading to massive bleeding. One factor that has been identified as causing increased acid production is the presence in the stomach of the bacterium *Helicobacter pylori*. Treatment with antibiotics to kill the bacteria can eliminate gastric ulcers in many patients.

In addition to these cell types, the epithelium of the stomach has many cells which produce a neutral or alkaline mucus. This protects the surface of the stomach from the damaging effects of hydrochloric acid.

22.5 Small intestine and accessory glands

Food passes through the pyloric sphincter from the stomach into the duodenum. In total the small intestine is 6 m long and its coiled tubes fill most of the abdominal cavity. In the duodenum the secretions of the liver and the pancreas are added to the stomach contents.

The pancreas

This is a gland which produces large numbers of digestive enzymes in a bicarbonate-rich fluid. The secretion of these enzymes is stimulated by the emptying of gastric contents into the duodenum. Most of the pancreatic enzymes are secreted as inactive precursors which are activated on entry into the GI tract. For example, the proteolytic enzyme trypsin is produced from the inactive precursor trypsinogen and itself then activates a number of other enzymes.

The liver

This vitally important organ has two key functional roles. Firstly it is a gland which is connected to the duodenum and which secretes *bile* through the *bile duct* into the GI tract. Prior to its secretion bile is stored in the *gall bladder*. Bile is a solution of bile salts which act to emulsify dietary lipids and are essential for their further digestion and absorption. Bile also contains the pigment *bilirubin*, which is derived from the breakdown of the haem component of red blood cells. The oxidation of bilirubin in the GI tract into a brownish pigment is what gives faeces their characteristic colour. Secondly, the liver has key metabolic roles. It is structured so that the blood returning from the GI tract filters past the walls of the cells of the liver, the *hepatocytes*, and these cells act to modify blood composition in the following ways.

- *Glucose regulation* – the liver is the primary site of glucose storage in the body and, following a meal, glucose is taken up into the liver and stored as glycogen. As glucose levels in the blood drop then the liver releases glucose to maintain a constant level.
- *Plasma protein synthesis* – the liver manufactures the majority of plasma proteins including albumin and the elements of the clotting cascade.
- *Detoxification* – the liver is responsible for taking potentially toxic substances, such as drugs and alcohol, out of the circulation and breaking them down into metabolites that can be excreted.
- *Lipid metabolism* – the liver is responsible for the storage and metabolism of fatty acids and triglycerides.

Once the stomach contents have been mixed with the secretions of the pancreas and liver then the resultant fluid (*chyme*) moves along the small intestine. The cells of the epithelium of the small intestine possess a surface coated with small finger-like projections called *microvilli*. This is known as a *brush border* and it increases the surface area of the cells vastly. The brush border is coated with a layer of enzymes which have been secreted by the epithelial cells themselves and which are responsible for the terminal digestion of the food. Thus, on this surface, disaccharides are converted to monosaccharides and dipeptides are converted to individual amino acids. Absorption of these final products of digestion is then achieved mostly through Na^+-coupled transport mechanisms (see Chapter 9) in the membrane of the epithelial cells (Fig. 22.3). The absorbed substances pass into the capillaries which run in the wall of the intestine. From there the blood travels to the liver before reentering the systemic circulation in the inferior vena cava.

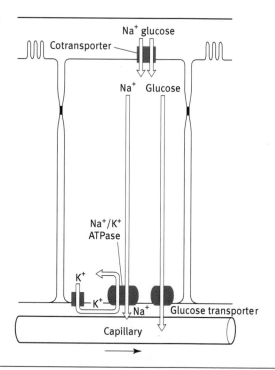

Figure 22.3. Absorption of glucose in the small intestine

22.6 Large intestine

The primary function of the upper parts of the small intestine is to digest and absorb proteins, carbohydrates and nucleic acids. As the small intestine joins the large intestine, the functional focus shifts to the absorption of lipids, water and electrolytes. As the chyme passes along the large intestine through the *colon* and into the *rectum*, water and electrolytes are absorbed so that the faeces become compacted. As the rectum, which lies at the end of the large intestine nearest to the anus, fills with faeces the mechanical stretching triggers a desire to relax the sphincter around the anal canal. The opening of this sphincter is under voluntary control. Once the sphincter is open contractions of the rectum force its contents out through the anal canal in the act of *defecation*.

22.7 Test yourself

The answers are given on p. 177.

Question 22.1
What are the three main layers of the wall of the gastrointestinal tube?

Question 22.2
What term describes the wave-like movements of the gut wall which propel food along the GI tract?

Question 22.3
What is the name given to the circular band of muscle that lies: (a) at the entrance to the stomach; (b) at the exit from the stomach into the small intestine?

Question 22.4
What substance is secreted by the liver into the duodenum and what is its function?

Question 22.5
What name is given to the finger-like projections on the luminal surface of epithelial cells of the small intestine and what is their function?

⬢23 The reproductive system

> **BASIC CONCEPTS:**
> Reproduction ensures the perpetuation of the species. It provides the mechanism whereby the male and female gametes (sperm and eggs) can develop and be brought together so that fertilisation and the subsequent development of an embryo can occur. The regulation of all stages of reproduction from the formation of gametes to the act of birth itself is under complex hormonal control and it is important to understand this in order to be able to understand how sexual function and fertility may be regulated.

23.1 Introduction

The function of the male and female reproductive systems is to ensure the survival of the species by generating male and female *gametes* (*spermatozoa* and *ova* respectively) and providing a means of these being brought together so that *fertilisation* may occur (see Chapter 11). Following fertilisation, the female reproductive system is responsible for supporting the developing embryo and its placenta until the baby has grown sufficiently to be born.

23.2 Male reproductive system

The male reproductive system (Fig. 23.1) comprises paired gonads (testes) which are each connected via the *epididymis* and the *vas deferens* to the *urethra*, a muscular tube which opens at the tip of the *penis*. The testes and epididymis are the sites of sperm production and maturation. This occurs optimally at 34°C, which is why they are located outside the abdominal cavity in a muscular sac known as the *scrotum*. The vas deferens runs from each testis to join with the urethra at the base of the bladder. The urethra conveys both urine and sperm to the exterior. During sexual intercourse sperm are released from the penis via the urethra in a process known as ejaculation. Ejaculated sperm are contained in about 5 ml of seminal fluid, which is derived from the secretions of accessory glands such as the prostate and seminal vesicles.

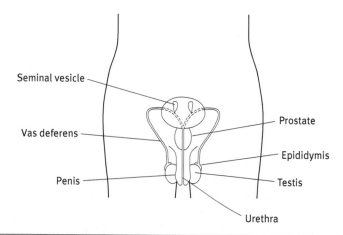

Seminal vesicle

Vas deferens

Penis

Prostate

Epididymis

Testis

Urethra

Figure 23.1. The male reproductive system

23.3 Spermatogenesis

Each testis contains an extensively coiled tube, the *seminiferous tubule*, within which sperm production takes place. Spermatozoa are derived initially from the mitotic division of primary germ cells known as *spermatogonia* which produces many *spermatocytes* which undergo meiotic division to produce haploid *spermatids*. The spermatids are embedded in *Sertoli cells* which support their differentiation into spermatozoa under the influence of FSH (see Chapter 18). Maturation of the spermatozoa from inactive to active cells takes place in the epididymis.

Within the connective tissue surrounding the seminiferous tubules are located *Leydig* cells which, under the influence of LH (see Chapter 18), are responsible for the production of the male sex hormone *testosterone*. Testosterone acts in the testis to promote spermatogenesis and is responsible for the development and maintenance of secondary sexual characteristics (e.g. beard growth) and sexual behaviour. Sperm are produced throughout the reproductive life of the male and, though sperm production drops gradually with age, sperm from a 90-year-old male are still capable of fertilising an ovum.

23.4 Female reproductive system

The female reproductive system (Fig. 23.2) comprises paired gonads (*ovaries*) which are linked via the *Fallopian tubes* to the *uterus*, and then through the *cervix* and *vagina* to the exterior. The ovaries lie within the abdominal cavity and the ova which they release have to travel a short distance within the abdominal cavity before being drawn into the Fallopian

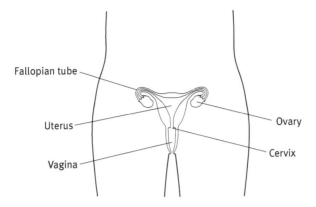

Figure 23.2. The female reproductive system

tubes by the beating of ciliated cells. The Fallopian tubes enter the uterus, which is a pear-shaped muscular sac whose lining is designed to support the fertilised ovum during its development into a baby. The uterus opens through a narrow aperture, the cervix, into the vagina which is a muscular canal opening to the outside. The vagina acts as a receptacle for the erect penis during copulation (sexual intercourse) and as the birth canal during *parturition* (the act of giving birth).

23.5   Oogenesis and the menstrual cycle

Female germ cells undergo extensive mitotic division in the foetal ovary. At birth, the ovaries contain about 1 million *primary oocytes*. These are cells arrested in the first stage of their meiotic division. Only 200–400 of these will ultimately be released as mature ova and be available for fertilisation – the rest will degenerate. A small number of these primary oocytes become surrounded by a layer of *granulosa* cells to form a *primary follicle*. Between birth and puberty these follicles are not able to mature appropriately. In a female who has passed puberty there is a monthly *menstrual* cycle (Fig. 23.3) in which a single follicle matures and releases its oocyte in a process known as *ovulation*. The oocyte together with protective layers which surround it is known as an *ovum*. This cycle is under hormonal regulation and cycles only cease during pregnancy and following the menopause, which occurs after the age of approximately 50 when no more follicles are available to mature.

Menstruation marks the end of one menstrual cycle and the beginning of the next and involves the shedding of the lining of the uterus along with a small amount of blood. Each menstrual cycle lasts approximately 28 days. During the first phase of the cycle, which begins with the onset of menstruation, raised levels of FSH (see Chapter 18) recruit a cohort of primary follicles

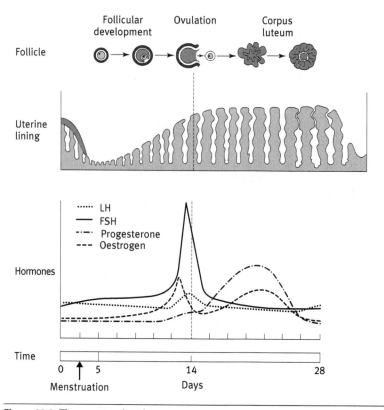

Figure 23.3. The menstrual cycle

which begin to mature. As they do so they secrete increasing levels of oestrogen which stimulates the lining of the uterus to begin to proliferate. As the follicles mature, one becomes gradually dominant and the others regress. Just before the midpoint of the cycle, at about 14 days, rapidly rising oestrogen levels trigger a surge in LH production which causes ovulation. The follicle, in response to LH, then transforms into a structure known as the *corpus luteum* which secretes the hormone progesterone. This stimulates further growth of the lining of the uterus. Towards the end of the menstrual cycle one of two things may happen:

- a successful fertilisation of the released ovum will have occurred and it will implant into the uterine lining;
- in the absence of fertilisation, the corpus luteum will cease hormone production and the uterine lining will degenerate and be shed, thus signalling the onset of menstruation and the start of a new cycle.

> **CLINICAL EXAMPLE: THE CONTRACEPTIVE PILL**
>
> The contraceptive pill contains a mixture of synthetic oestrogens and progesterones. These act to provide negative feedback to the pituitary and hypothalamus where they suppress the production of FSH and LH. This prevents follicle development and ovulation. However, the synthetic hormones do allow growth of the lining of the uterus, and so stopping the drugs for 7 days out of every 28 allows menstruation to occur normally.

23.6 Copulation and fertilisation

Copulation involves the insertion of the erect penis into the vagina for the purposes of triggering ejaculation and the delivery of seminal fluid containing 150 to 600 million sperm into the female reproductive tract. The sperm then use their flagella to drive themselves through the cervix into the uterus and into the Fallopian tubes where fertilisation normally occurs. Only about 100 sperm complete this journey successfully and on the way they undergo a process known as *capacitation*, which renders them capable of penetrating and fertilising the ovum. When the sperm encounters an ovum it passes through the glycoprotein coat which surrounds the ovum and its head fuses with the cell membrane of the oocyte. This triggers the oocyte to alter the structure of its wall to prevent fertilisation by other sperm and to complete its meiotic division. The haploid sperm and oocyte nuclei then fuse to create a new diploid cell. *Conception* has occurred.

23.7 Implantation and pregnancy

Following fertilisation, the ovum begins to divide and soon creates a small ball of cells known as a *blastocyst*. This will eventually give rise to both the embryo and the placenta. In its early stages the blastocyst begins to secrete the hormone human chorionic gonadotrophin (hCG) which signals the corpus luteum not to regress and thus prevents the uterine lining from being shed. Detection of hCG is used in pregnancy-testing kits. The blastocyst lies in the uterine cavity for about 72 hours before attaching to and invading the uterine wall in the process of implantation. As the cells in the blastocyst continue to divide they form a hollow ball of cells. The cells on the outer wall of this ball will form the fetal part of the placenta and the inner cell mass will develop into the fetus. As the developing fetus burrows further it induces cells of the uterine wall to participate in forming the placenta. The placenta consists of blood vessels from the fetus in close proximity with blood from the mother. This allows exchange of oxygen, carbon dioxide, nutrients and waste products between the fetal and maternal circulations.

The fetus will normally spend about 9 months in the uterus. During the first three months (*trimester*) most of the vital organs are formed, and the remaining six months are largely devoted to growth.

23.8 Birth

As the result of a number of factors, including endocrine signals and mechanical stretching of the uterus, the woman will go into labour. This involves rhythmic contractions of the uterus stimulated by the release of bursts of oxytocin from the posterior pituitary which increase in frequency as labour progresses. These contractions are accompanied by hormonally regulated relaxation and dilation of the cervix and vagina to allow the passage of the baby to the outside world.

> **CLINICAL EXAMPLE: ECTOPIC PREGNANCY**
>
> An ectopic pregnancy is one in which implantation occurs in a site other than the uterus. Ectopic pregnancies commonly occur in the Fallopian tubes but can also occur on the outside of abdominal or pelvic organs if the ovum escapes into the abdominal cavity. It can be particularly dangerous because the fetus is developing in a site which is not adapted to receive it. As the fetus grows it can cause rupture of the wall of the organ into which it is embedded and lead to excessive bleeding. A woman suffering from a ruptured ectopic pregnancy will complain of sudden abdominal pain and show signs of excessive blood loss such as increased heart rate and a pale, clammy skin.

23.9 Test yourself

The answers are given on p. 177.

Question 23.1
What name is given to: (a) the male gonads; (b) the muscular sac in which they are contained?

Question 23.2
Down which structures must the ova pass after leaving the ovary in order to reach the uterus?

Question 23.3
Which hormones stimulate the lining of the uterus to proliferate and are: (a) secreted by maturing primary follicles early in the cycle; (b) secreted by the corpus luteum later in the cycle?

Question 23.4
What hormone is secreted by the blastocyst after fertilisation and can form the basis of a pregnancy test?

Question 23.5
Which pituitary hormone stimulates the uterus to contract during labour?

24 The urinary system

BASIC CONCEPTS:
The maintenance of body fluid and electrolyte levels is vital to ensure functioning of all cells, but particularly the excitable tissues. For example, small fluctuations in plasma potassium levels can lead to heart failure and paralysis. The kidney is responsible for this regulation. This is achieved by a process of filtering the fluids and electrolytes from plasma and then reabsorbing sufficient quantities of each into the blood to maintain appropriate plasma levels. Reabsorption in the kidneys is under hormonal control. The kidney is also involved in the excretion of the waste nitrogen formed from the breakdown of protein. Kidney failure is fatal and so it is vital to understand the functioning of this complex organ.

24.1 The kidney

The kidney is one of the main homeostatic organs in the body. Chapter 18 explains how hormones secreted by the kidney control red blood cell production and peripheral resistance. The kidney is primarily responsible for *osmoregulation* – the regulation of fluid and electrolyte balance within the body. In association with this function, the kidney is also responsible for *excretion* – the removal of waste products. These include *urea* (waste nitrogen) and other metabolites. The output of the kidney is *urine*, a concentrated solution of urea. This is passed to the bladder for storage before it is voided to the exterior via the urethra (Fig. 24.1).

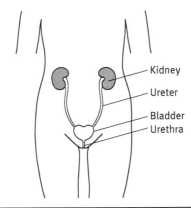

Kidney
Ureter
Bladder
Urethra

Figure 24.1. The urinary system

The kidneys are paired organs situated on the posterior wall of the abdomen, where they are protected from damage by the lower ribs. They receive a blood supply from the aorta through the *renal arteries* and return filtered blood to the circulation through *renal veins*. A cross-section through a kidney (Fig. 24.2) shows three main areas – an outer *cortex*, an inner *medulla* and the renal pelvis where urine is collected and drains into the *ureter* and thence to the *bladder*.

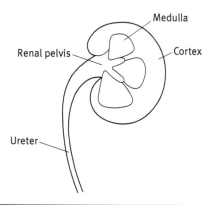

Figure 24.2. Cross-section of the kidney

The basic functional unit of the kidney is the *nephron* (Fig. 24.3) which is a tubular structure closed at one end and open at the other. Each kidney contains between one and two million nephrons.

At the closed end of each nephron is *Bowman's capsule*. This is invaginated by a knot of capillaries, the *glomerulus*. Most glomeruli are located in the renal cortex. In this region, filtration of the plasma takes place. Plasma proteins do not pass through Bowman's capsule. Approximately 6 litres of plasma is filtered by the kidneys each hour. The filtrate is iso-osmolar with plasma and contains water, electrolytes, urea, glucose and amino acids. The function of the nephrons is to reclaim the required amounts of useful materials whilst reducing the 6 l/hr of filtrate to approximately 60 ml/hr of a concentrated urea solution.

This process begins in the proximal tubule with the active reabsorption of glucose, amino acids, NaCl and other electrolytes. The movement of these solutes generates an osmotic force which draws water along with them. About 80% of the filtered water is reabsorbed here, along with all the glucose and amino acids. Reabsorption of sodium in the proximal tubule is under the control of the hormone angiotensin II and this is the major site for regulation of blood volume and pressure (see Chapter 20).

The filtrate then passes into the descending limb of the *loop of Henle*. This part of the nephron dips into the renal medulla. The descending and ascending limbs of the loop of Henle operate a countercurrent system which

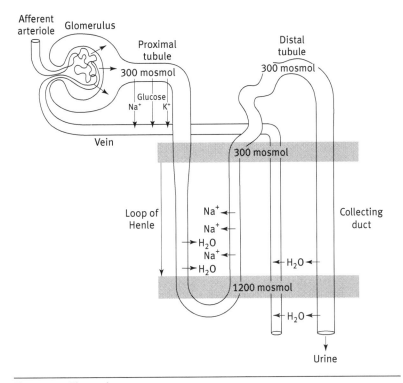

Figure 24.3. The nephron

results in an osmolarity gradient being established between the top of the loop and the bottom. At the bottom of the loop of Henle, the tubular and interstitial fluids have an osmolarity of 1200 mosmol – about four times that of plasma. The gradient is established because the ascending limb is impermeable to water but actively pumps sodium into the descending limb, thus increasing its solute concentration.

Fluid from the ascending limb then enters the distal convoluted tubule where more sodium is actively pumped out along with water. The removal of sodium here is under the control of the hormone aldosterone which is produced by the adrenal gland in response to decreased plasma Na$^+$ ion concentration or in response to dehydration. Increased sodium reabsorption also increases water re-uptake.

The final removal of water takes place in the collecting ducts which dip back down through the renal medulla towards the renal pelvis. As they pass through the renal medulla they run through the area of high osmolarity created at the bottom of the loop of Henle. The presence of specific water channels (*aquaporins*) in the cells of the epithelium of the collecting duct allows rapid transfer of water out of the duct and into the interstitial fluid. The aquaporins are less permeable to the movement of urea than of water, with the consequence that urea becomes concentrated within the urine.

The number of aquaporin channels present in the collecting duct is regulated by antidiuretic hormone (ADH) which is produced in the hypothalamus in response to decreasing water content of blood (rising plasma osmolarity). This triggers the insertion of more aquaporin channels and thus allows more water to be reabsorbed from the filtrate. In the event of a rise in plasma volume (for instance after drinking large volumes of water) ADH secretion is reduced, aquaporin channels are withdrawn and more water flows through the collecting ducts into the ureter and down to the bladder.

24.2   The bladder

This is a distensible muscular sac with a wall containing a thick layer of smooth muscle. The bladder is capable of storing up to 1 litre of urine. The bladder is connected to the exterior via the *urethra*. Stretching the bladder wall generates the urge to pass urine, but relaxation of muscular sphincters around the urethra is needed before emptying can take place. The neural control of these sphincters is through a combination of voluntary and involuntary mechanisms.

CLINICAL EXAMPLE: RENAL FAILURE

There are many causes of renal failure but one of the most common occurs in response to prolonged high blood pressure. This initially causes damage to the filtration barrier in the kidney glomerulus and leads to the excretion of protein and glucose in the urine. The presence of these substances in urine can prevent water being reabsorbed and lead to the patient producing large quantities of dilute urine. Eventually the damaged nephrons die and then the patient will not have enough functional kidney tissue to enable them to clear all the toxic metabolites from their blood. In this situation an artificial kidney (a dialysis machine) is used to help clear the patient's blood of metabolites until they can receive a kidney transplant.

24.3   Test yourself

The answers are given on p. 177.

Question 24.1
What is the basic functional unit of the kidney?

Question 24.2
What is the primary function of Bowman's capsule and the glomerulus?

Question 24.3
Where in the kidney does most reabsorption of glucose, sodium and other electrolytes take place?

Question 24.4
Which hormone acts to regulate reabsorption of sodium in the distal convoluted tubule?

Question 24.5
What effect does antidiuretic hormone (ADH) have on the kidney?

25 The immune system

> **BASIC CONCEPTS:**
> The body is under constant threat from microorganisms in the external
> environment and these seek to use the body's resources to ensure their
> own survival. Such infections are potentially fatal and the survival of the
> individual depends on being able to destroy or neutralise these infective
> agents. The immune system functions to protect the body against infection
> by recognising invading microorganisms and destroying them. It does this
> by employing a wide range of molecules and cell types which interact to
> produce a coordinated response. Occasionally this response can be
> targeted against the body itself and can be severely debilitating. An
> understanding of the interactions between the cells of the immune system
> and their targets is important in appreciating how microorganisms interact
> with the host to cause disease. Knowledge of the immune system can also
> help to control both infectious and other non-infectious diseases.

25.1 Immune responses to infection

The body's system of defence against infection is the immune system. The
immune system has a large number of different weapons in its armoury to
enable it to deal with different types of infection arising in different parts of
the body. These weapons consist of various cells and molecules that are able
to attack infectious agents.

Barriers to infection

The first line of defence against infection is provided by the barriers that exist
between our bodies and their surroundings. The skin is impervious to most
infectious agents and is protected by acidic secretions that are hostile to
most microorganisms. Mucosal surfaces, such as those lining the respiratory
tract, are protected by a layer of mucus, a highly viscous secretion that
effectively traps microorganisms and can be moved along the mucosal tract
by cilia on epithelial cells. Secretions such as tears and digestive juices
contain the enzyme lysozyme, which attacks bacterial cell walls.

Molecules and cells of the immune system

If an infectious agent manages to breach these barrier defences and gain access to the body it will then come into contact with further molecules and cells of the immune system. *Phagocytes* are cells that can ingest microorganisms and then destroy them. There are two main types of phagocyte – *neutrophilic granulocytes* and *macrophages*. Granulocytes are the predominant type of white blood cell. They circulate round the body and home in on sites of infection. Macrophages are found in certain tissues, such as the lungs, where they provide a resident population of immune cells. Both granulocytes and macrophages produce a range of molecules that can attack any ingested microorganisms. Microorganisms are internalised into a vesicle, which then fuses with lysosomes of the cell, exposing the microorganism to their various enzymes. A range of molecules found in blood and tissue fluids can also attack infectious agents. One of the most important of these is *complement*. Complement is a group of plasma proteins which, in the presence of infection, acts to promote phagocytosis. It does this by binding to the surface of microorganisms and marking them as foreign for ingestion by phagocytes (Fig. 25.1). Complement can also act to promote the process of inflammation.

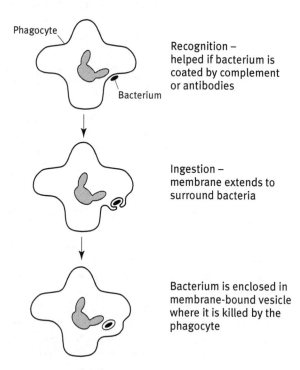

Figure 25.1. The stages of phagocytosis

25.2 Inflammation

When an infectious agent enters a tissue of the body it is important that the cells and molecules of the immune system are brought into the site of infection so they can start to launch an attack. This is brought about by the process of *inflammation*. Inflammation involves changes in the blood vessels that supply a tissue making the vessel walls more permeable. This allows circulating cells and fluid to leave the bloodstream and enter the tissues. Interaction of microorganisms with tissue cells induces the production of molecules, which in turn act on the blood vessels to bring about the increased permeability. One of the principal cells found at site of inflammation is the granulocyte, which can help dispose of the infectious agent by phagocytosis.

25.3 Lymphocytes and the specific immune response

The defences against infection that have been described so far are present in the body, even before an infection comes along, and form part of the *innate* immune system.

The innate defences are fast-acting and can help to keep an infection at bay while the body develops its second line of defence – a *specific immune response*. Specific immune responses are produced by *lymphocytes* which circulate round the body, moving between the blood and tissues and seeking out infection. There are two main types of lymphocytes, *B cells* and *T cells*. Both T and B cells have receptors on their surface that are highly specific for individual molecular structures of different infectious agents. Each cell has only one type of receptor and can therefore only recognise and respond to one type of infection. The structures that are recognised by the T and B cell receptors are called *antigens*. Antigens can include proteins, polysaccharides and other molecules produced by microorganisms. The number of different antigens that can be recognised by the specific immune system is enormous, meaning that the body is able to defend itself against any microorganism it might encounter. This is because of the very large number of different T and B cell receptors, which has been estimated to be in the region of 10^9.

T and B cells have different mechanisms for fighting infection. When a B cell comes into contact with its specific antigen, it differentiates into plasma cells which secrete proteins called *antibodies*. Antibodies are found in the blood, tissue fluids and mucosal secretions. They are able to bind to antigens with a high degree of specificity. This can help fight infection in several ways – antibodies binding to antigens on the surface of a virus can block that virus from infecting a body cell; antibodies can also promote phagocytosis in a similar way to complement, by binding to microorganisms and marking them as foreign.

While antibodies can act against microbes that occur in extracellular locations within the body, T cells are particularly good at recognising intracellular infections such as viruses. *Cytotoxic T cells* can recognise cells of the body that contain a virus infection and kill these cells, so limiting the replication and spread of the virus. *Helper T cells* act by secreting proteins called cytokines that promote the activities of various other cells of the immune system. For example, helper T cells are important in activating B cells to allow them to secrete antibodies, and in interacting with macrophages to up-regulate their killing mechanisms. The functions of T and B lymphocytes are summarised in the table below.

Type of cell	Function
B lymphocyte	Gives rise to antibody-secreting plasma cells
Helper T cell	Secretes cytokines that help other cells – for example activate macrophages; help B cells become antibody-secreting cells
Cytotoxic T cell	Kills virus-infected body cells

25.4 Diseases of the immune system

HIV and immunodeficiency

The importance of the immune system in defending against infection is clearly seen in individuals whose immune function is compromised. Such people are described as being *immunodeficient*. Immunodeficiency can be due to a number of causes including malnutrition, certain types of cancer and some drug treatments. Infection can also cause immunodeficiency, and the best illustration of this is the HIV virus.

HIV is an RNA retrovirus. It infects cells of the immune system including helper T cells and macrophages. After infection the RNA genome is converted into DNA by the viral enzyme *reverse transcriptase*. This DNA can integrate itself into the DNA of the host cells and, in this form, the virus can remain latent for years, with the infected person showing no symptoms. After a period of time the virus becomes activated and this leads to the depletion of helper T cells. When numbers of helper T cells fall below a critical level the patient starts to succumb to a range of infections. Some of these infections involve microorganisms that pose no threat to healthy individuals and these are described as *opportunistic* infections. If untreated, the HIV-infected patient will eventually die, normally as a result of infection.

Treatment against HIV involves drugs that act against the virus. Two types of drug are commonly used – inhibitors of reverse transcriptase and inhibitors of the viral protease. The viral genes are transcribed and translated into a large protein which is then cleaved by a viral protease to give rise to the proteins of the virus. If this cleavage is blocked by protease inhibitors then virus production is prevented.

Allergy and autoimmunity

Although the immune system is crucial in protecting against infection, unfortunate consequences for the host can arise when immune responses are made inappropriately. Two examples of this are diseases arising from *allergy* and *autoimmunity*.

In allergy, an immune response occurs against a substance that would not normally cause problems for the host. For example, hay fever sufferers produce antibodies against grass pollen. These antibodies cause inflammation in the upper respiratory tract leading to the symptoms of the disease – runny nose, sneezing, etc. Some allergies are more serious. Asthma can be caused by allergic reactions lower down the respiratory tract leading to problems with breathing that can be life-threatening. The most serious type of allergic reaction is seen to antigens that enter the systemic circulation. This can occur with some food antigens, such as those derived from peanuts. In this instance the inflammation resulting from the allergy can affect blood vessels throughout the whole body and can very quickly lead to collapse and death if treatment is not administered rapidly.

In autoimmunity, the immune response starts to attack the body's own cells and tissues instead of being directed against an infectious agent. Under normal circumstances lymphocytes that have receptors for the body's own molecules (self-antigens) are eliminated or rendered non-responsive – a process known as immunological tolerance. Autoimmune disease arises when something goes wrong with this process. Autoimmune diseases can affect a range of tissues of the body and include conditions such as diabetes, rheumatoid arthritis and multiple sclerosis.

25.5 Test yourself

The answers are given on p. 177.

Question 25.1
Identify two aspects of the body's defence mechanisms which are promoted by complement.

Question 25.2
What name is given to a molecular structure recognised by the receptors of T and B lymphocytes?

Question 25.3
Which cells of the immune system recognise and kill virus-infected cells?

Question 25.4
Which type of phagocytic cells circulate in the blood and enter sites of inflammation?

Question 25.5
What molecules secreted by plasma cells bind to a specific antigen?

26 The musculoskeletal system

> **BASIC CONCEPTS:**
> The action of muscles on the bony skeleton generates movement at joints. Control of movement is achieved through the action of motor neurons which are regulated by local reflexes in the spinal cord and by descending control from the brain. Bone is a connective tissue which acts as an important store of calcium, the plasma levels of which are under hormonal control. Adequate levels of calcium in the blood are essential for normal functioning of nerve and muscle and so it is important to understand these regulatory mechanisms.

26.1 Introduction

Locomotion is one of the primary attributes of animals. In most vertebrates, movement of the limbs is achieved by the action of muscles pulling on a jointed skeleton. The skeleton is composed predominantly of the connective tissue, *bone*. Bones articulate smoothly with each other at *synovial joints* which act as low-friction shock absorbers due to the presence of *cartilage* on their surfaces. Muscles acting on the skeleton are known as skeletal muscles and are controlled by somatic nerves which arise from the brainstem and spinal cord (see Chapter 19).

26.2 Bone

Bone is the major component of the skeleton. It is a strong, rigid connective tissue. Like all other connective tissues in the body, bone is a living tissue and depends on adequate supplies of nutrients and oxygen to stay alive. Connective tissues (see Chapter 15) are composed of three main elements – cells and fibres embedded in a protein, and proteoglycan matrix. Bone differs from other connective tissues only in the presence within this matrix of the crystalline mineral hydroxyapatite – $Ca_{10}(PO_4)_6.(OH)_2$ – which gives bone its rigidity.

Bone is constantly being remodelled, with old bone being eaten away by cells known as *osteoclasts*. These are multinucleated phagocytic cells derived from blood-borne precursors. Their activity can be stimulated by increased levels of thyroid and parathyroid hormone and inhibited by oestrogen. New bone is laid down in the spaces which osteoclasts create by cells known as *osteoblasts*. These cells secrete the molecules which make up the extracellular matrix of bone and can increase their activity in response to growth hormone and insulin-like growth factor. Following deposition of the matrix the process of mineralisation, during which hydroxyapatite crystals grow in the matrix, is largely dependent on plasma calcium levels.

In the adult skeleton the activity of osteoclasts and osteoblasts needs to be kept in balance in order to maintain bone density. This is achieved via a complex pattern of hormonal regulation which is partially related to plasma calcium levels since bone is the principal calcium store in the body. There is also direct coupling between osteoblast and osteoclast activity.

26.3   Regulation of calcium levels

Plasma calcium levels are tightly regulated. Variations in plasma calcium above or below the normal range can have profound effects on excitable tissues. Plasma calcium levels depend on three main processes:

- amount of dietary calcium absorbed from the gut;
- amount of calcium deposited in or released from bone;
- amount of calcium excreted by the kidney.

Alteration of any one of these processes will have consequences for the others. Thus a decrease in calcium uptake from the gut will lead to an increase in calcium release from bone and contribute to a demineralisation of the bone matrix.

Plasma calcium levels are detected by cells in the parathyroid gland, kidney, brain and other organs. The two main substances which control calcium levels are parathyroid hormone (see Chapter 18) and vitamin D.

CLINICAL EXAMPLE: OSTEOPOROSIS

Osteoporosis is the loss of bone tissue due to excessive activity of osteoclasts. It is a particular problem in post-menopausal women because of the sudden loss of the inhibitory effect of oestrogen on osteoclasts. In a sufferer from osteoporosis, bones become more brittle and susceptible to fracture. Typically this is seen in the fractured hips which occur as the result of seemingly innocuous falls in the elderly.

Vitamin D can be derived from the diet and the most important dietary sources are fish, liver and milk. However, the major source of vitamin D for most individuals is from synthesis by cells in the skin as a direct result of the action of sunlight. The actions of vitamin D are to:

- increase absorption of calcium from the gut;
- increase removal of bone by osteoclasts.

26.4 Cartilage

In addition to bone, the skeleton also contains cartilage. This is found on the surfaces of bones at joints and plays an important role during the growth of bones. Cartilage is a connective tissue which contains a dense extracellular matrix containing collagen and sulphated proteoglycans. Sulphate ions attract Na^+ ions into the matrix which in turn draw and hold in water by osmosis. The presence of water in the matrix makes cartilage resistant to compression and it is thus ideally structured to resist forces transmitted through joints during locomotion.

26.5 The skeleton

The bones in the skeleton can be classified as *long* bones or flat bones. A typical long bone is the femur which is located in the thigh. This has a long, straight hollow shaft containing a cavity filled with *bone marrow*. Bone marrow is the site of production of red and white blood cells. At both ends of the femur are heads which form a movable synovial joint with the adjacent bone at either the hip or the knee. A typical flat bone is the frontal bone (forehead), one of the bones which make up the skull. Flat bones tend to be protective in function and their joints with adjacent bones are often rigid. They also contain a marrow cavity.

26.6 Synovial joints

Synovial joints enable adjacent bones to move relative to each other (Fig. 26.1). For example, between the femur in the thigh and the tibia in the lower leg is the knee joint which allows the movements of flexion (bending) and extension (straightening). The joint is surrounded by a capsule of tough connective tissue which is lined by the synovial membrane. This secretes fluid into the joint cavity which acts both as a lubricant for the joint surfaces and to convey nutrients to the cartilage covering the ends of the bones. The cartilage on the bones provides a low-friction surface to facilitate movement and acts as a shock absorber to minimise the jarring effects of walking or running. The stability of synovial joints is increased by the presence of strong ligaments which surround the joint and by the muscles which act on the joint. This is why physiotherapy for the muscles around a joint is important to athletes recovering from joint damage.

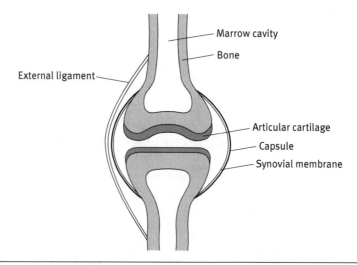

Figure 26.1. A synovial joint

26.7 Muscles and locomotion

The structure and function of muscle fibres is covered in Chapter 16. Individual muscles, such as the quadriceps muscles in the front of the thigh, are made up of many muscle fibres embedded in a connective tissue matrix. Muscles which move a joint are often inserted directly into the bone above the joint and then give rise to a tendon which crosses the joint to be moved and is then inserted into the bone below the joint. During walking or running, the knee joint undergoes alternate flexion and extension movements. This is achieved by a controlled pattern of contraction and relaxation of the muscles in the front and back of the thigh. The *extensor* muscles in the front of the thigh straighten the leg as an individual pushes off from the ground and the *flexor* muscles bend the knee as the free leg swings forward. These alternate movements are controlled by sets of neurons in the spinal cord which act as *locomotor pattern generators* and are not subject to conscious regulation.

Individual motor neurons within the spinal cord send axons to innervate one or more muscle fibres. The number of muscle fibres controlled by a single neuron is known as the *motor unit* and varies depending on the level of control required. Thus, muscles which control coarse movements, such as those in the thigh, will have large motor units (up to 200 fibres) while in the fingers precision of movement is achieved by having a one-to-one relationship between a motor neuron and a muscle fibre.

In order for muscle contraction to be regulated effectively it is important that the central nervous system is constantly aware of the state of contraction of the muscle. This information is provided by specialised sensory fibres located within muscles and joints and is known as *proprioception*. An example of proprioception in action can be seen in the knee-jerk reflex, in which the tendon at the front of the knee joint is stretched by tapping it with a hammer and the thigh muscles immediately straighten the leg to counteract the stretch (Fig. 26.2). The neural basis for this is that sensory nerve endings in the muscle are stimulated by the tendon being stretched and their axons run back into the spinal cord where they directly excite the motor neurons which innervate the quadriceps muscle. They also indirectly cause relaxation of the muscles in the back of the thigh. Reflexes like this operate continuously during locomotion to counteract unexpected movements of the joint, such as those produced when individuals travel over rough ground.

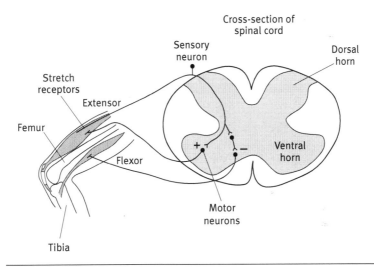

Figure 26.2. The knee-jerk reflex

During locomotion, individual muscles are not under direct conscious control but deep nuclei in the brain send signals via the cerebral cortex to the spinal cord which allow the local pattern generators to produce movement. The smooth execution of movement is due to the actions of the cerebellum (see Chapter 19). The brain can also exert voluntary control over individual muscles by sending signals directly from neurons in the cerebral cortex to individual motor neurons in the spinal cord.

26.8 Test yourself

The answers are given on p. 177.

Question 26.1
Which cells are responsible for the resorption of bone?

Question 26.2
What are the two main actions of vitamin D?

Question 26.3
What lies in the centre of the shaft of a long bone?

Question 26.4
What is a motor unit?

Question 26.5
In the knee-jerk reflex, which muscles contract in response to tapping the tendon at the front of the knee joint?

Answers to 'test yourself' questions

Answer 1.1
On the inside of the molecule

Answer 1.2
Intracellular fluid, interstitial fluid, plasma

Answer 1.3
(a) Na^+; (b) K^+

Answer 1.4
Carbon

Answer 1.5
Protein

Answer 2.1
Amino acids

Answer 2.2
Amino (NH_2) and carboxyl (COOH)

Answer 2.3
The sequence of amino acids

Answer 2.4
(a) On the outside of the helix; (b) every 3.6 residues; (c) hydrogen bonds

Answer 2.5
(a) Homodimer; (b) Heterodimer

Answer 3.1
$C_6H_{12}O_6$

Answer 3.2
Glycosidic bond

Answer 3.3
Amylase

Answer 3.4
Glycosyl transferase

Answer 3.5
Glycogen

Answer 4.1
Saturated

Answer 4.2
Glycerol

Answer 4.3
Two

Answer 4.4
Amphipathic

Answer 4.5
Adipose tissue

Answer 5.1
Deoxyribonucleic acid

Answer 5.2
Adenine (A), thymine (T), guanine (G) and cytosine (C). A pairs with T; C pairs with G

Answer 5.3
A codon

Answer 5.4
Promoter

Answer 5.5
Transfer RNA (tRNA)

Answer 6.1
Lysosomes

Answer 6.2
In mitochondria

Answer 6.3
Mitochondria

Answer 6.4
Actin

Answer 6.5
Ribosomes

Answer 7.1
Peptidoglycan

Answer 7.2
Reverse transcription

Answer 7.3
True – this is the bacteriophages, sometimes called phages for short

Answer 7.4
Chitin

Answer 7.5
A unicellular eukaryotic parasite or protozoa of the genus *Plasmodium*

Answer 8.1
Pyruvate

Answer 8.2
Adenosine triphosphate

Answer 8.3
In mitochondria

Answer 8.4
Oxidative phosphorylation

Answer 8.5
Lactic acid

Answer 9.1
3 osmolar

Answer 9.2
True

Answer 9.3
Na+

Answer 9.4
Sodium/potassium ATPase

Answer 9.5
No, because it can freely diffuse through the cell membrane

Answer 10.1
G1, S and G2

Answer 10.2
The mitotic spindle

Answer 10.3
The centromere

Answer 10.4
Telomeres

Answer 10.5
Cyclin-dependent kinases

Answer 11.1
Gametes (eggs and sperm)

Answer 11.2
Meiosis

Answer 11.3
Chiasmata

Answer 11.4
Zona pellucida

Answer 11.5
Embryonic stem cells

Answer 12.1
Homozygous

Answer 12.2
Insertion

Answer 12.3
They are likely to be linked – i.e. on the same chromosome

Answer 12.4
(a) XX; (b) XY

Answer 12.5
Selective advantage

Answer 13.1
Because the mitochondrial DNA is inherited from the mother

Answer 13.2
Transport of chloride ions across cell membranes

Answer 13.3
True

Answer 13.4
Haemophilia arises due to defects in the clotting cascade, which involves a number of proteins encoded by different genes

Answer 13.5
Zero, because the gene is on the X chromosome, and sons always inherit this from the mother

Answer 14.1
One

Answer 14.2
Stratified squamous

Answer 14.3
Basal lamina

Answer 14.4
Tight junctions

Answer 14.5
Desmosomes

Answer 15.1
Glycosaminoglycans

Answer 15.2
Collagen

Answer 15.3
Chondroblasts

Answer 15.4
Chondrocytes

Answer 15.5
To promote inflammation

Answer 16.1
K+

Answer 16.2
Actin and myosin

Answer 16.3
Acetylcholine

Answer 16.4
The cell body, the axon and dendrites

Answer 16.5
To increase the rate of conduction by providing electrical insulation for the axon

Answer 17.1
Insulin

Answer 17.2
Glucagon

Answer 17.3
The hypothalamus

Answer 17.4
The blood flow to the skin is reduced due to narrowing of the peripheral blood vessels

Answer 17.5
Brown fat

Answer 18.1
Steroids

Answer 18.2
Adrenalin

Answer 18.3
GTP

Answer 18.4
Cortisol

Answer 18.5
(a) Thyroid stimulating hormone (TSH); growth hormone (GH); adrenocorticotrophic hormone (ACTH); luteinising hormone (LH); follicle stimulating hormone (FSH); and prolactin (PRL)
(b) Antidiuretic hormone (ADH); oxytocin

Answer 19.1
Motor neurons

Answer 19.2
Frontal, parietal, occipital and temporal

Answer 19.3
The somatic division

Answer 19.4
(a) Noradrenalin; (b) acetylcholine

Answer 19.5
The vagus nerve

Answer 20.1
Fibrin

Answer 20.2
(a) Systole; (b) diastole

Answer 20.3
(a) Tricuspid valve; (b) bicuspid or mitral valve

Answer 20.4
(a) Left atrium; (b) right atrium

Answer 20.5
Heart rate, stroke volume and peripheral resistance

Answer 21.1
The pleural membrane

Answer 21.2
Ciliated columnar cells and goblet cells

Answer 21.3
The diaphragm

Answer 21.4
Squamous alveolar cells and capillary endothelial cells

Answer 21.5
The partial pressure of CO_2 in the blood

Answer 22.1
The mucosa, the submucosa and the external muscle coat

Answer 22.2
Peristalsis

Answer 22.3
(a) Oesophageal sphincter; (b) pyloric sphincter

Answer 22.4
Bile – this contains bile salts which serve to emulsify lipids

Answer 22.5
Microvilli – these greatly increase the surface area for absorption

Answer 23.1
(a) Testes; (b) scrotum

Answer 23.2
Fallopian tubes

Answer 23.3
(a) Oestrogen; (b) progesterone

Answer 23.4
Human chorionic gonadotrophin (hCG)

Answer 23.5
Oxytocin

Answer 24.1
Nephron

Answer 24.2
Filtration

Answer 24.3
In the proximal convoluted tubule

Answer 24.4
Aldosterone

Answer 24.5
It causes the insertion of water channels into the collecting ducts to increase reabsorption of water

Answer 25.1
Phagocytosis; inflammation

Answer 25.2
An antigen

Answer 25.3
Cytotoxic T cells

Answer 25.4
Granulocytes – more specifically, neutrophilic granulocytes

Answer 25.5
Antibodies

Answer 26.1
Osteoclasts

Answer 26.2
It increases absorption of calcium from the gut and increases osteoclast activity

Answer 26.3
The marrow cavity containing bone marrow

Answer 26.4
The number of muscle fibres innervated by one motor neuron

Answer 26.5
The extensor muscles of the thigh

Glossary

absolute refractory period: the period during which a section of the axonal membrane cannot be depolarised following the passage of an action potential

acetylation: the addition of an acetyl group $-COCH_3$ to another molecule

acetylcholine: an excitatory neurotransmitter molecule used at various synapses in the central nervous system, in the neuromuscular junction and by the parasympathetic nervous system

acetylcholinesterase: an enzyme located in the synaptic cleft of cholinergic synapses which acts to break down acetylcholine and thus to terminate its action

acrosome: a structure at the tip of a sperm head that is responsible for enabling the sperm to penetrate the layers surrounding the egg and initiating fertilisation

actin: a globular protein which can be assembled into a microfilament; it plays an important role in cell motility

action potential: a wave of depolarisation which travels away from the cell body of a neuron along the axon and which triggers transmitter release at the axon terminal

active transport: a mechanism by which substances are moved across a cell membrane against their concentration gradient using energy derived from ATP

adenine: a purine base found in DNA and RNA; it is paired with thymine in DNA and uracil in RNA

adenosine triphosphate (ATP): the major source of usable energy in cell metabolism; composed of adenine, ribose, and three phosphate groups

adenylate cyclase: an enzyme located in the cell membrane which can be activated by the α-subunit of G-proteins to convert ATP into cyclic AMP

adipocyte: the main cell type of fat; capable of storing triglycerides in a single large globule within its cytoplasm

adipose tissue: a connective tissue, made up largely of adipocytes, which acts as the major site for storage of triglycerides within the body

adrenaline: a hormone (also known as epinephrine) released by the medulla of the adrenal gland during the 'flight or fight' reaction; it produces a variety of effects around the body including increasing heart rate

adrenocorticotrophic hormone (ACTH): a hormone produced by corticotrophs in the anterior pituitary; it acts on the adrenal cortex to stimulate the production and release of corticosteroids

aggrecan: a large aggregate of proteoglycan which together with hyaluronic acid forms the basis of the extracellular matrix in many connective tissues

albumin: the most abundant and most negatively charged plasma protein; it is produced in the liver and acts as a carrier molecule for substances being transported in the bloodstream; it is a major contributor to plasma oncotic pressure

aldosterone: a hormone produced by the adrenal cortex which acts on the distal convoluted tubule in the kidney where it regulates sodium reabsorption

allele: one of the alternative forms of the same functional gene; alleles occupy the same position (locus) on homologous chromosomes

alpha helix: a secondary structure in proteins in which the polypeptide chain twists upon itself in a highly regular helical manner; each turn of the helix represents 3.6 amino acid residues

alveoli (*singular alveolus*): in the lungs the alveoli are thin-walled air-filled sacs which lie at the end of the respiratory tree and in which gaseous exchange takes place; the term alveoli is also used to describe clusters of secretory cells in exocrine glands

amino acid: an organic acid containing the amino group (NH_2) and serving as a building block for proteins

amphipathic: a molecule which has both hydrophilic and hydrophobic moeities

amylase: a digestive enzyme which can hydrolyse starch

amylose: an unbranched form of starch in which glucose residues have an α-1,4 linkage

anaerobic respiration: the breakdown of organic molecules and the release of energy in the absence of oxygen

anaphase: a stage of mitosis during which the sister chromatids separate and move away from each other

angiotensin: a hormone which acts to raise blood pressure; produced by the action of renin on angiotensinogen; the product angiotensin I is converted to the active form angiotensin II by the action of angiotensin converting enzyme (ACE) found on the walls of capillaries

anion: a negatively charged atom or group

antibiotic: one of a range of drugs which has the effect of preventing bacterial growth either by killing bacteria or preventing their replication

antibody: a protein produced by plasma cells in response to a foreign organism or molecule; by binding to the foreign organism or molecule it inactivates it or marks it for destruction

anticodon: a sequence of three nucleotides in a transfer RNA molecule which hydrogen bonds to the complementary codon in a messenger RNA

antidiuretic hormone (ADH): a hormone produced in the hypothalamus and released from the posterior pituitary; it acts on the collecting duct in the kidney to promote water reabsorption; also known as vasopressin

antigen: a substance which can be recognised by antibodies and T cell receptors to provoke an immune response

aquaporin: one of a family of membrane channels which permit the rapid movement of water across cell membranes

artery: a vessel which carries blood away from the heart

asexual reproduction: reproduction which does not involve the fusion of gametes; it produces an organism identical to the parent

ATP synthase: an enzyme complex on the inner mitochondrial membrane which utilises the energy from proton flow to convert ADP to ATP

atrium: a chamber of the heart which receives venous blood and passes it on to a ventricle

atrio-ventricular (A-V) node: specialised tissue in the heart wall which initiates ventricular contraction by transmitting impulses along the A-V bundle

autoimmunity: a process in which the immune system recognises components of the body as foreign and attacks them

autonomic nervous system: a division of the peripheral nervous system which is divided into sympathetic and parasympathetic branches; it innervates smooth muscle and glandular tissue

autosome: chromosome other than the sex chromosomes; in human cells this refers to chromosomes 1–22

axon: a long thin process of a neuron; it transmits an action potential away from the cell body; it ends at a synapse or neuromuscular junction where neurotransmitter is released

B cell (B lymphocyte): a cell of the immune system which gives rise to an antibody-producing plasma cell

bacteria (*singular bacterium*): prokaryotic organisms surrounded by a rigid proteoglycan cell wall

bacteriophage: a virus that specifically infects bacteria

baroreceptor: sensory structure located in major arteries capable of detecting changes in blood pressure

base: in the context of DNA or RNA; the purine or pyrimidine component of nucleotides

basal lamina: a thin molecular matrix secreted from the base of epithelial cells which provides linkage between them and underlying

connective tissue basal transcription factors: a group of proteins which must bind to the promotor region of a gene before transcription can occur

beta cell: a cell in the pancreatic islets which secretes insulin in response to rising glucose levels

beta sheet: a secondary structure found in proteins in which the amino acids associate with each other through hydrogen bonds to form a rigid, folded sheet-like structure

bilirubin: a pigment released during the breakdown of red blood cells; derived from the iron-containing haem group of haemoglobin

binary fission: a process in which a single cell produces two identical daughter cells; it is a term most often used to describe the process whereby a prokaryote replicates its single DNA strand and then divides into two daughter cells

blastocyst: an early embryonic stage prior to implantation consisting of a small ball of cells

Bowman's capsule: the closed end of the nephron into which the knot of capillaries known as the glomerulus is invaginated; the site of plasma filtration

brown fat: a specialised adipose tissue which can generate heat and is important in thermoregulation in infants

brush border: the microvilli on the surface of absorptive cells in the gut and elsewhere

calmodulin: a protein activated by calcium binding which modifies the activity of target enzymes

capacitation: a process which the head of a sperm undergoes in the female genital tract in preparation for fertilisation

capillary: a thin-walled vessel which joins the arterial to the venous sides of the circulation; the site where substances move between plasma and interstitial fluid

capsid: the outer protein coat enclosing the nucleic acid genome of a virus

carbohydrate: molecules, such as sugar and starch, with the general formula $(CH_2O)_n$

cation: a positively charged atom or group

cell cycle: the stages through which a cell passes whilst undergoing mitotic division; can be divided into interphase, during which DNA synthesis occurs, and mitotic (M)-phase, during which the cell divides

cell membrane: a phospholipid bilayer surrounding all cells

cellulase: an enzyme which can break down cellulose; often found in the digestive tracts of herbivorous animals but not in humans

cellulose: an unbranched polymer of glucose linked by β-1,4 glycosidic bonds which forms an important component of plant cell walls

central dogma: the concept that information flows from DNA to RNA to protein

centromere: the structure which, in a replicated chromosome prior to mitosis, holds together the two sister chromatids

centrosome: an area in the cell cytoplasm responsible for stabilising the microtubular component of the cytoskeleton either in interphase or during spindle formation in mitosis

checkpoint: a point during the cell cycle at which a decision is made whether or not to proceed to the next stage; progress through a checkpoint is regulated by cyclin-dependent kinases

chiasmata (*singular chiasma*): the points of connection at which parts of homologous chromosomes are exchanged during meiosis

chief cell: enzyme-secreting cell in the glands of the stomach

chitin: a long chain polysaccharide comprising repeated sub-units of *N*-acetylglucosamine; it is the main component of the cell walls of fungi

cholesterol: a lipid molecule based on a structure containing four linked hydrocarbon rings and a short hydrocarbon tail; it plays an important role in determining the fluidity and permeability of plasma membranes and forms the start point for the synthesis of steroid hormones

chondroblast: a cell responsible for the deposition of the extracellular matrix in newly forming cartilage

chondrocyte: a mature cell located within the matrix of cartilage responsible for the maintenance of the extracellular matrix

chromatid: following DNA replication (in S-phase) the two identical copies of each of the 46 chromosomes are known as sister chromatids and are joined to each other at the centromere

chromosome: a strand of DNA containing a number of genes together with associated proteins; human cells contain 46 chromosomes in 23 pairs; in each pair one chromosome is derived from the mother and one from the father

cilium (*plural cilia*): an extension of the apical surface of epithelial cells containing a specialised array of microtubules which allows it to move backwards and forwards in a wave-like fashion

citric acid cycle: a metabolic pathway located within the mitochondrion which oxidises acetyl groups derived from the breakdown of organic molecules to generate NADH and CO_2

clone: a cell or organism genetically identical to the cell or organism from which it was derived

codon: a sequence of three nucleotides within a messenger RNA molecule that specifies a particular amino acid or the start or stop of translation

collagen: a triple helical protein made up from three -chains which can assemble into fibrils and fibres; it provides the strength and flexibility of connective tissues

complement: a group of plasma proteins which when activated can protect against infection by binding to the surface of infecting microorganisms and promoting their phagocytosis and by promoting inflammation

compliance: in pulmonary physiology describes the ease with which a lung can be expanded

condensation reaction: a chemical reaction in which water is removed

conjugation: a mechanism by which genetic material (usually in the form of a plasmid) is transferred from one bacterium to another through direct cell contact; it allows for horizontal gene transfer

corpus luteum: a structure in the ovary derived from a post-ovulatory follicle which maintains the uterine lining in readiness for implantation of the fertilised ovum

cortisol: a hormone produced by the cortex of the adrenal gland which has stimulatory effects on many metabolic processes and is chronically elevated in stress

co-transporter: a membrane protein which facilitates the transport of one substance up its concentration gradient by linking it with the transport of another substance (usually Na^+) down its concentration gradient; also called symport

covalent bond: a chemical bond formed between two atoms in which they share a pair of electrons

C-terminus: The end of a polypeptide chain at which the amino acid has a free carboxylic acid group

cyclic AMP: adenosine monophosphate in a form in which the single phosphate links back to the ribose sugar; it is an important second messenger in many signalling pathways

cyclin-dependent kinase: a protein kinase involved in the regulation of the cell cycle; cyclin-dependent kinases are active when combined with regulatory proteins known as cyclins

cystic fibrosis transmembrane conductance regulator (CFTR): a protein, located in the apical membrane of many epithelial cells, which acts as an ATP-regulated chloride channel; it also has an inhibitory effect on co-located epithelial sodium channels

cytokine: a type of signal molecule which is secreted from one cell and acts on another

cytokinesis: the final stage of mitosis during which the two daughter cells separate

cytoplasm: that part of the internal fluid compartment of the cell which is bounded by the plasma membrane but not contained within the nucleus or other cell organelles

cytosine: a pyrimidine base found in DNA and RNA; it pairs with guanine

cytoskeleton: the internal protein framework of the cell made up of microfilaments, intermediate filaments and microtubules

deletion: a mutation in which one or more nucleotides are lost from the DNA strand

dendrite: a branched extension from the cell body of a neuron onto which synapses are made by axons of other neurons

dense body: a point of attachment for the sarcomeres of smooth muscle cells which is equivalent to the Z-line of striated muscle

depolarisation: a change in the membrane potential of a cell such that the inside becomes less negative with respect to the outside; usually caused by the entry of cations into the cell

desmosome: a cell junction at which protein bridges provide strong mechanical linkage between two adjacent cells

diastole: that phase of the cardiac cycle during which the ventricles relax and the atria contract

differentiation: the process by which a cell acquires specialised structure and function via the expression of a specific set of genes

diffusion: the movement of molecules and small particles from an area of higher concentration to an area of lower concentration by random, thermally driven movements

dimer: a molecule such as a protein comprising two associated sub-units

disulphide bridge: a covalent bond formed between two -SH groups; occurs between cysteine residues in proteins

DNA: deoxyribonucleic acid (DNA); a polymer of deoxyribonucleotides; this is the genetic material for all life forms and adopts a double helical structure; the sequence of bases in the nucleotides provides the code for the synthesis of proteins

domain: a part of a protein with a specific structure and function

dominant: a gene whose phenotype is expressed in both the homozygous and heterozygous state

electrical potential: a voltage difference between the inside and the outside of a cell membrane caused by an uneven distribution of charge

electrical syncytium: a group of cells linked by gap junctions through which depolarisation can spread extremely rapidly

endocytosis: a process whereby membrane-bound vesicles are formed at the cell surface and taken into the cytoplasm, leading to the internalisation of extracellular substances

endoplasmic reticulum: a series of membrane bound spaces within the cytoplasm of cells which act as the sites of protein synthesis and other metabolic processes

endothelium: the simple squamous epithelial lining of blood vessels and the chambers of the heart

enhancer: a region of DNA, often located some distance away from a gene, to which regulatory proteins may bind and increase transcription of that gene

enteric nervous system: a collection of neurons located in and regulating the activity of the gastro-intestinal tract

enzyme: a protein that catalyses a specific chemical reaction

epidermis: the outer epithelial layer of the skin

erythrocyte: a red blood cell

erythropoietin: a hormone produced in the kidney in response to low oxygen levels in the blood which stimulates the production of erythrocytes

eukaryote: an organism based on cells which possess a distinct nucleus and cytoplasm

exocytosis: the process by which membrane-bound vesicles fuse with the cell membrane and release their contents into the extracellular space

exon: a section of a eukaryotic gene that encodes the amino acid sequence for part of a protein

extracellular fluid: the fluid which lies outside cells; in the body it is divided into interstitial fluid which is in direct contact with cells and plasma which is contained within blood vessels

extracellular matrix: the extracellular component of connective tissues; it contains proteoglycans, glycoproteins and water; collagen and elastic fibres are embedded within it

facilitated diffusion: a process in which the rate of movement of a substance across a cell membrane and down its concentration gradient is enhanced by the presence in the membrane of transporter proteins or channels

fatty acid: a molecule that has a carboxylic acid group attached to a long hydrocarbon tail; used as a major energy source during metabolism; forms the hydrophobic tails of membrane phospholipids

fertilisation: the process which begins when a sperm and an egg come together; it leads to their fusion and the initiation of the development of the embryo

fibrin: a protein formed from fibrinogen by the action of thrombin; it is the major protein component of blood clots

fibroblast: a cell found in fibrous connective tissues such as tendon and responsible for the secretion of the extracellular matrix; fibroblasts proliferate in response to injury causing fibrosis or scarring

fibrocyte: a mature cell of fibrous connective tissues such as tendon; responsible for the maintenance of the extracellular matrix

flagellum: a long protrusion on a cell which propels the cell through a fluid medium by its beating; e.g. the sperm tail

follicle stimulating hormone (FSH): a hormone secreted by gonadotrophs in the anterior pituitary; in females FSH stimulates ovarian follicles to mature; in males FSH promotes spermatogenesis

gamete: a haploid cell (carrying only one set of chromosomes) formed as part of sexual reproduction by meiotic division; a sperm or an egg

gametogenesis: the process by which gametes are formed

gap junction: a cell-to-cell junction which permits the passage of ions and small molecules from one cell to another

gene: a region of DNA within a chromosome that codes for a specific protein

genetic code: the rules that determine which triplet of nucleotides (codons) in DNA code for which amino acid in protein synthesis

genetic locus: the position occupied by a gene on a chromosome

genotype: the full set of genes carried by an individual cell or organism

gestation period: the time between fertilisation and birth

glia: supporting cells in the nervous system; responsible for maintenance of a constant micro-environment, myelination and reaction to injury

glomerulus: the knot of capillaries which lies within Bowman's capsule in a nephron; may also be used to describe the capsule plus the capillaries

glucagon: a hormone produced by -cells in the pancreas in response to falling glucose levels; it stimulates glycogen breakdown in liver and muscle

glucose: a six carbon sugar that has the formula $C_6H_{12}O_6$; a major source of metabolic energy which can be stored as glycogen (animals) or starch (plants)

glyceride: a lipid formed when the carboxylic acid group on a fatty acid reacts with glycerol

glycogen: a branched polysaccharide of glucose; a major energy store in liver and muscle cells

glycolipid: a membrane lipid in which the polar head group has been glycosylated

glycolysis: a metabolic pathway in which glucose is broken down into two three-carbon pyruvate molecules with the production of a small amount of ATP and NADH; takes place in the cytoplasm

glycosylation: the addition of sugars to another biological molecule

glycoprotein: a protein which has been glycosylated

glycosaminoglycan: a high molecular weight polysaccharide containing repeated subunits of amino and carboxylic acid sugars which is prevalent in the matrix of connective tissues

glycosidase: an enzyme which catalyses the breakdown of a glycosidic bond

glycosidic bond: a covalent bond formed between two monosaccharides by a condensation reaction

Golgi apparatus: an organelle consisting of stacked membrane sacs in which newly synthesised proteins are sorted and packaged into membrane-bound vesicles for transport to the cell surface or to lysosomes

G-protein: a trimeric intracellular GTP-binding protein associated with the cytoplasmic domain of a transmembrane receptor; following receptor-ligand interaction, the G-protein sub-units diffuse away from the receptor and initiate the production of second messengers

Gram stain: a stain which binds to peptidoglycan in the cell walls of certain bacteria; it is used as the basis of rapid laboratory classification of bacteria into Gram-positive and Gram-negative types

grey matter: that part of the central nervous system containing neuronal cell bodies, dendrites, synapses and glial cells

growth factors: cell signalling molecules which stimulate growth and differentiation of their target cells or organs

growth hormone: a hormone produced by the somatotrophs in the anterior pituitary; it has multiple trophic effects on its target tissues; necessary for normal development of an individual

guanine: a purine base found in DNA and RNA; it pairs with cytosine

haemoglobin: a tetrameric protein found in red blood cells and responsible for the transport of oxygen from the lungs to the tissues of the body

haploid: carrying only one half of the normal chromosomal complement of a cell; i.e. a single copy of each chromosome

helicase: an enzyme which unwinds the DNA double helix prior to DNA replication

hepatocyte: the basic cell type of the liver; responsible for production of plasma proteins, glucose uptake and storage, detoxification and bile production

heterodimer: a molecule comprising two non-identical sub-units

heterozygous: in diploid organisms, having two different alleles at the same locus of homologous chromosomes

hexose: a sugar, such as glucose, which has six carbon atoms per molecule

histone: a protein component of eukaryotic chromosomes around which the DNA strand is coiled

homeostasis: the maintenance of a constant internal environment

homodimer: a molecule containing two identical sub-units

homologous chromosome: one of two copies of a specific chromosome in a diploid cell; one copy is inherited from the father and one from the mother

homozygous: in diploid organisms, having identical alleles at the same locus of homologous chromosomes

hormone: a signal molecule released from cells at one site and transported through the blood to exert a specific effect on its target cells at a different site

human chorionic gonadotrophin: a hormone released by the placenta which prevents regression of the corpus luteum; detection of its presence in urine can be used in pregnancy tests

hydrogen-bond: a bond formed between a hydrogen atom in one molecule and an electronegative atom (e.g. O or N) in another molecule or another part of the same molecule

hydrophilic: a molecule or part of a molecule which can readily interact with or dissolve in water; 'water-loving'

hydrophobic: a molecule or part of a molecule which is unable to interact with or dissolve in water; 'water-hating'

hydrophobic interaction: a repulsive interaction between a molecule and water; the coming together of hydrophobic groups because of their exclusion by water

hydroxyapatite: crystalline calcium phosphate; the mineral component of bone

hyperosmolar: describes a solution or fluid compartment having a higher osmolarity than another

hyperpolarisation: movement of the resting membrane potential to a more negative value than normal

hypertonic: describes a solution which because of its high effective osmolarity would cause water to move out of cells placed in it

hypha (plural: hyphae): a long filamentous multicellular structure of which most fungi are constructed; it comprises a tubular outer wall made of chitin which contains along its length multiple cells separated by septa; the cells are in cytoplasmic continuity through large pores in the septa

hypo-osmolar: describes a solution or fluid compartment having a lower osmolarity than another

hypotonic: describes a solution which because of its low effective osmolarity would cause water to move into cells placed in it

immunoglobulins: proteins produced by the immune system which protect against infection; antibodies

inflammation: a tissue response to insult or injury characterised by increased blood flow, localised oedema and infiltration of tissues by cells of the immune system

inspiration: the act of breathing in

insulin: a hormone produced by β-cells of the pancreas in response to increased plasma glucose; it stimulates glucose uptake and storage in cells of muscle and liver

intercalated disc: the junction between two adjacent cardiac muscle cells; it contains alternating desmosomes and gap junctions

intermediate filaments: strong intracellular filaments assembled from short rod-like proteins; the component of the cytoskeleton which provides mechanical stability e.g. keratin

interphase: the period in the cell cycle between cell divisions; comprises G1, S and G2 phases

interstitial fluid: the component of extracellular fluid which directly bathes the surface of cells

intracellular fluid: the 60% of body water which lies within cells and forms the solvent for cytoplasmic and nuclear contents

intron: a section of DNA within a gene which does not code directly for a polypeptide sequence

ionic bond: a chemical bond resulting from the attraction between oppositely charged ions

iso-osmolar: describes a solution or fluid compartment having the same osmolarity as another

isotonic: describes a solution which causes no net movement of water into or out of cells placed in it

keratin: a tough intermediate filament protein; found in high levels in skin, nails and hair

lectin: a protein which recognises and binds to specific carbohydrate side chains present on molecules such as glycoproteins and glycolipids

leptin: a hormone produced by cells in adipose tissue which acts to suppress appetite

Leydig cells: cells in the testis responsible for the production of testesterone

ligand: a molecule such as a hormone or neurotransmitter that binds to a specific site on a receptor protein

ligase: an enzyme which joins together DNA molecules

linkage: the tendency of certain genes to be inherited together as a result of their proximity on a chromosome

lipid: an organic molecule which is insoluble in water

luteinising hormone: a hormone produced by gonadotrophs in the anterior pituitary which in the female triggers ovulation and formation of the corpus luteum; in the male it stimulates Leydig cells to produce testosterone

lymphocyte: a cell of the immune system involved in the immune response

lysosome: a cytoplasmic organelle with a low internal pH containing hydrolytic enzymes involved in the breakdown and recycling of cellular macromolecules

lysozyme: an enzyme which attacks peptidoglycan in bacterial cell walls

macrophage: a cell specialised for the phagocytosis of particulate matter

mast cell: a cell found in connective tissues which is involved in local inflammatory responses

meiosis: cell division which leads to the production of four haploid gametes from a single diploid parent cell

messenger RNA (mRNA): a molecule produced as a copy of the DNA coding strand during transcription; each mRNA codes for a specific protein

metaphase: a stage of mitosis in which the chromosomes arrange themselves in the centre of the cell, equidistant from each pole of the spindle

microfilament: a thin intracellular protein filament such as actin or myosin; the part of the cytoskeleton which is primarily involved in cellular motility

microtubule: a tubular cytoskeletal component made from sub-units of the protein tubulin; it can be rapidly assembled and disassembled; important in intracellular transport and in the mitotic spindle

microvilli (*singular microvillus*): small finger-like membrane projections specialised for absorption which increase the surface area of epithelial cells

mitochondria (*singular mitochondrion*): cytoplasmic organelles which contain the enzymes necessary for the citric acid cycle and oxidative phosphorylation; they are responsible for the production of ATP during aerobic respiration

mitosis: the part of the cell cycle during which the cell divides and the sister chromatids that have been produced as a result of DNA replication are separated and distributed equally between the two daughter cells

mitotic spindle: a specialised array of microtubules formed during mitosis which is responsible for separation of the sister chromatids during anaphase and elongation of the cell during telophase

molecular chaperone: a molecule which assists in the folding or intracellular transport of a polypeptide

monosaccharide: a simple sugar molecule with the general formula $(CH_2O)_n$

motor neuron: a neuron whose axon innervates one or more muscle fibres

motor unit: the total number of muscle fibres innervated by the axonal branches of one neuron

mucin: a large glycoprotein containing a high proportion of O-linked oligosaccharides; a major constituent of mucus

mucus: a sticky glandular secretion containing high levels of mucin; it protects and lubricates epithelial surfaces

mutation: a random change in the nucleotide sequence of DNA within a chromosome

myocyte: a muscle cell

myelination: the process of forming a myelin sheath

myelin sheath: a layer of electrical insulation around an axon; formed from the cell membranes of glial cells which wind around the axon

myofibrils: highly organised bundles of the microfilaments actin and myosin found within the cytoplasm of muscle cells; myofibrils are made up of repeated units known as sarcomeres and shorten during muscle contraction

myosin: a protein which can break down ATP to drive its movement relative to actin microfilaments; one of the key components of muscle cells

Na^+/K^+ ATPase: a membrane pump found in many cells which uses energy derived from ATP to pump three sodium ions out of the cell and two potassium ions into the cell

negative feedback loop: a process in homeostatic or metabolic regulation by which the product of a reaction feeds back to inhibit that reaction and thus slows down or terminates its own production

nephron: the basic functional unit of the kidney; responsible for filtration of plasma and the production of urine; has a central role in osmoregulation

neuromuscular junction: a specialised point of contact between the terminal of the axon of a motor neuron and a muscle fibre at which the release of acetylcholine triggers muscle contraction

neuron: an excitable cell within the nervous system; comprises a cell body, axon and dendrites

neurotransmitter: a chemical released from an axon at a synapse or neuromuscular junction which binds to a receptor on the post-synaptic target cell and triggers a response

neutrophilic granulocyte: a phagocytic white blood cell; important in defence against infection

nicotine adenine dinucleotide (NAD^+): a carrier molecule which accepts a hydride ion (H^-) from a donor molecule to produce NADH; participates in energy production from the breakdown of glucose and other molecules

node of Ranvier: a small area of an axon between myelinated sections where the axonal membrane is in direct contact with interstitial fluid and where depolarisation events occur during saltatory conduction of an action potential

noradrenaline: a neurotransmitter (also known as norepinephrine) used at various synapses in the central nervous system and in the periphery by the sympathetic nervous system

N-terminus: the end of a polypeptide chain at which the amino acid has a free amino group

nucleic acid: a polymer assembled from nucleotides; the two principal nucleic acids are deoxyribonucleic acid (DNA) and ribonucleic acid (RNA)

nucleosome: the basic structural unit of the eukaryotic chromosome comprising DNA coiled around a core of histone proteins

nucleotide: a molecule containing a purine or pyrimidine base, a pentose sugar (ribose or deoxyribose) and a phosphate group

nucleus: a cellular organelle characteristic of eukaryotic cells; has a double membrane with pores surrounding the genetic material of the cell; contains the enzymes and proteins necessary for DNA transcription

oedema: the accumulation of fluid in the interstitial spaces between cells; swelling

oesophagus: a muscular tube used to convey food from the mouth to the stomach

oestrogens: a class of steroid hormones (e.g. oestradiol) produced in the ovary which initiate growth of the uterine lining during the menstrual cycle and are responsible for the development of secondary sexual characteristics in the female

oligodendroglia: glial cells which myelinate axons in the central nervous system

oncotic pressure: the osmotic force which acts to draw water back into capillaries from the interstitial fluid

oocyte: a female gamete

oogenesis: the process of production of primary oocytes within the ovary

operator: a region of an operon which binds a repressor or inducer to regulate the expression of the genes in the operon

operon: a sequence of DNA found in bacteria containing an operator, a promotor and a structural gene or genes transcribed as a single unit

opportunistic infection: an infection in which microorganisms that are not normally pathogenic cause disease, usually in people with reduced immune function (e.g. as a result of HIV infection)

opsonisation: the coating of an infectious agent with molecules such as antibodies to enable recognition by phagocytes

organelle: a structure or compartment within a eukaryotic cell which is specialised to carry out a specific function

osmolality: a measure of the number of particles of solute per mass of solvent; expressed as osmol/kg

osmolarity: a measure of the number of particles dissolved within a solution; determined by the molarity of the solute multiplied by the number of particles into which it dissociates; expressed as osmol/l.

osmoregulation: a homeostatic mechanism by which the osmolarity of body fluids is maintained

osmosis: the bulk transfer of water across a semi-permeable membrane from an area of low solute concentration to one of higher solute concentration; the membrane must be less permeable to the solute than to water

osteoblast: a cell responsible for secreting the extracellular matrix of developing bone and for initiating mineralisation

osteoclast: a multinucleated phagocytic cell which resorbs bone and releases calcium into the blood-stream

osteocyte: a mature cell of bone responsible for the maintenance of the extracellular matrix

ovum: an oocyte together with the protective layers that surround it

oxidative phosphorylation: a process located on the inner membrane of mitochondria in which energy derived from the breakdown of organic molecules is used to generate ATP

oxytocin: a hormone produced in the hypothalamus and released from the posterior pituitary which stimulates contraction of the uterus during birth and milk release during suckling

parasympathetic nervous system: a division of the autonomic nervous system which slows down the heart and promotes digestive processes

parathyroid hormone: a hormone secreted by the parathyroid glands in response to decreasing plasma calcium, which stimulates the release of calcium from bone

parietal cell: a cell found in glands in the stomach; secretes HCl and vitamin B_{12} intrinsic factor

partial pressure: a measure of the amount of dissolved gas in a solution; equivalent to the proportion of atmospheric pressure that would be provided by that gas if it were present in the atmosphere at a level sufficient to equilibrate with the solution

parturition: the act of giving birth

passive transport: the movement of a molecule across the cell membrane down its concentration gradient; occurs by diffusion or facilitated diffusion and does not require energy

pathogen: an organism which can cause disease

pentose: a sugar with the general formula $(CH_2O)_5$; e.g. ribose

peptide bond: a bond formed by a condensation reaction between the carboxylic acid group of one amino acid and the amino group of another

peptidoglycan: a polymer which is a component of the cell wall of bacteria; it comprises glycosaminoglycan molecules cross-linked by short peptide chains

peptidyl transferase: an activity within ribosomes which catalyses the formation of peptide bonds between the NH2 group of an incoming amino acid on a tRNA molecule and the free COOH group of the preceding amino acid

peripheral resistance: the resistance to blood flow provided predominantly by small arterioles in the peripheral circulation

peristalsis: the wave-like contraction of a muscular tube such as the gut to move matter along its length

phagocyte: a cell specialised to perform phagocytosis; important in defence against infection e.g. a macrophage

phagocytosis: ingestion of particulate matter by a cell

phenotype: the observable characteristics of an organism 3'-5'

phosphodiester bond: a bond in nucleic acids in which the –OH on the 3rd carbon of one sugar is covalently bound to the phosphate group, which in turn forms a bond with the –OH on the 5th carbon of the next sugar

phospholipid: an amphipathic lipid found in cell membranes; many phospholipids are based on glycerol to which are linked two fatty acid chains, a phosphate group and a polar head molecule

phosphorylation: the addition of a phosphate group ($-PO_4$) to a molecule

pilus (plural: pili): a tubular extension from the wall of a bacterium which can have multiple functions; some types of pili initiate contact between adjacent bacteria at the start of the process of conjugation, while other pili generate movement of the bacteria relative to its substrate

placenta: the organ that develops during pregnancy to provide the interface between the mother and developing foetus

plasma: the fluid component of the blood

plasma cell: a cell that secretes antibody; derived from differentiation of a B lymphocyte

platelet: a cellular fragment found in large numbers in the blood; important for clotting

point mutation: a change in a gene involving a single base

polar: describes a molecule or group with an uneven distribution of charge

polygenic: describes a characteristic that is dependent upon multiple genes

polymer: a molecule, linear or branched, composed of multiple sub-units

polymerase: an enzyme that catalyses the addition of sub-units to form a polymer; e.g. DNA polymerase, RNA polymerase

polymorphic: in relation to a gene, refers to having multiple variants (alleles) within the population

polypeptide: a polymer of amino acids; polypeptides comprise a single chain whereas proteins may contain one or more polypeptide chains

polyunsaturated: a type of fatty acid containing multiple double bonds

postsynaptic membrane: the dendritic membrane adjacent to the synaptic cleft in which neurotransmitter receptors are localised

presynaptic membrane: the axonal membrane adjacent to the synaptic cleft from which neurotransmitter is released

prokaryote: an organism based on cells that lack a nucleus; e.g. bacteria

prolactin: a hormone produced by lactotrophs in the anterior pituitary; it initiates and helps maintain milk production

prometaphase: a stage of mitosis in which the nuclear membrane starts to break down so the chromosomes can become attached to the mitotic spindle

promoter: a sequence upstream of a gene to which RNA polymerase binds to initiate transcription

prophase: a stage of mitosis in which the chromosomes condense and the mitotic spindle starts to form

proprioception: the sense of the position of the body, or parts of the body, in space

protease: an enzyme which is able to break down a protein into short peptide sequences by hydrolysing peptide bonds

protein: a polymer of amino acids comprising one or more polypeptide chains; one of a large group of molecules found in all living organisms

protein kinase: an enzyme that adds phosphate groups from ATP to specific amino acid residues in protein molecules

proteoglycan: a macromolecule consisting of a core protein with covalently-linked, negatively-charged sulphated glycosaminoglycans; e.g. aggrecan, a major component of the extracellular matrix

protozoa: a single-celled eukaryotic organism; some species can cause infectious disease, e.g. malaria

purine: a molecule based on a nitrogen-containing 6-membered ring fused to a second 5-membered ring found in the nucleotides of DNA and RNA; includes adenine and guanine

pyrimidine: a molecule based on a nitrogen-containing 6-membered ring found in the nucleotides of DNA and RNA; includes thymine, cytosine and uracil

pyruvate: a three-carbon breakdown product formed from glucose by glycolysis

reading frame: the division of a nucleotide sequence into consecutive triplets which may code for specific amino acids; thus a sequence such as ACGGTACTA could be read as ACG GTA CTA or CGG TAC or GGT ACT depending on whether the reading frame starts at the first, second or third nucleotide

receptor: a molecule that binds specifically to another molecule (a ligand) normally with some biological consequence

recessive: a gene whose phenotype is only expressed in the homozygous state

recombination: a process in which two strands of nucleic acid break and become rejoined in a new relationship; occurs during the pairing of homologous chromosomes in meiosis

relative refractory period: a period following the absolute refractory period during which the membrane is slightly hyperpolarised and more difficult to depolarise again

renin: an enzyme secreted by the kidney that catalyses the production of angiotensin from an inactive precursor, leading to an increase in blood pressure

repressor: a protein that binds to DNA and reduces the transcription of a gene or genes

residual volume: that volume of air which is left in the lungs after expiration

resting membrane potential: the difference in electrical potential between the inside and outside of a cell membrane when the cell is in an unstimulated state

retrovirus: a virus with an RNA genome which uses reverse transcriptase to synthesise DNA; e.g. human immunodeficiency virus (HIV)

reverse transcriptase: an enzyme that catalyses the synthesis of DNA on an RNA template

ribosomal RNA: RNA that is part of the ribosome

ribosome: a subcellular particle comprising RNA and protein that catalyses protein synthesis

ribozyme: RNA with catalytic activity

ribonucleic acid (RNA): a polymer of ribonucleotides; occurs as messenger (m)RNA, transfer (t)RNA and ribosomal (r)RNA

saltatory conduction: a process in which an action potential jumps from one node of Ranvier to the next, allowing an increased conduction velocity

sarcomeres: repeated units that make up the myofibrils of striated muscle

Schwann cell: a glial cell which myelinates axons in the peripheral nervous system

seminal fluid: a secretion in which spermatozoa are ejaculated

sensory neuron: a neuron that transmits impulses from the periphery to the central nervous system

Sertoli cell: a cell within the testis that supports the production of spermatozoa

sex chromosome: one of the pair of chromosomes that determine the sex of an individual; X and Y in humans

sexual reproduction: a type of reproduction involving the mixing of genetic material from two parents to produce offspring that are genetically different from the parents and each other

smooth endoplasmic reticulum: endoplasmic reticulum that is not associated with ribosomes; plays a role in lipid synthesis, intracellular Ca^{2+} storage and detoxification processes

solute: a substance dissolved in solution

spermatid: an immature form of the spermatozoon

spermatozoon: a male gamete

spinal cord: part of the central nervous system lying within the vertebral canal

spinal nerve: a nerve originating from the spinal cord

stem cell: an undifferentiated cell which has the potential to divide and produce a wide range of differentiated cell types

steroid hormone: one of a family of hormones structurally related to cholesterol that acts via intracellular receptors to regulate gene expression

stroke volume: the volume of blood expelled from a ventricle as it contracts

surfactant: a lipid substance secreted by type II cells in the alveolus which reduces the alveolar surface tension, making the lungs easier to inflate

sympathetic nervous system: a division of the autonomic nervous system which controls the 'fight or flight' reaction

synapse: a structure found between the axon of one neuron and dendrites of another, across which nerve impulses can be transmitted

systole: that phase during the cardiac cycle in which the ventricles contract and the atria relax

T cell (T lymphocyte): a lymphocyte that undergoes maturation in the thymus and recognises antigens present on cells of the body

telomere: the end of a chromosome

telophase: a stage of mitosis in which a nuclear envelope is formed around the two separated sets of chromosomes prior to cell division

testosterone: a steroid hormone produced by the testis in the male and involved in the development and maintenance of secondary sexual characteristics

tetramer: a molecule made up of four sub-units e.g. haemoglobin

thermoreceptors: receptors that detect changes in temperature

thrombin: an enzyme that converts fibrinogen to fibrin during the formation of blood clots

thymine: a pyrimidine base found in DNA; it pairs with adenine

thyroid hormone: a hormone produced by the thyroid gland that stimulates metabolic activity

thyroid stimulating hormone: a hormone produced by thyrotrophs in the anterior pituitary; it stimulates the thyroid gland to produce thyroid hormone

tidal volume: the volume of air that is drawn into the lungs during inspiration

tight junction: a junction between epithelial cells where the adjacent cell membranes are held together very closely so as to prevent the passage of most substances from one side of an epithelial sheet to the other

tonicity: describes the effective osmolarity of a solution; an indication of the effect a solution is likely to have on cells placed in it

transcellular fluid: extracellular fluid that is not plasma or interstitial fluid; e.g. fluid in the bladder or the gut

transcription: the process by which a complementary strand of RNA is synthesised on a DNA template

transcription factor: a protein that regulates the transcription of a gene or genes

translation: the process by which a protein is synthesised on a strand of mRNA

transfer RNA (tRNA): a type of RNA molecule that carries the anticodon for lining up amino acids with codons on messenger RNA; each amino acid has a specific tRNA molecule

triglyceride: a derivative of glycerol with three fatty acids attached

trisomy: an abnormality in which three rather than two copies of a chromosome are present

tropomyosin: a filamentous protein associated with actin filaments in striated muscle; it blocks the actin domains to which myosin may bind

troponin: a protein associated with actin filaments in striated muscle

trypsin: a proteolytic enzyme secreted by the pancreas

trypsinogen: the inactive precursor of trypsin

t-tubules: membrane invaginations that run into the centre of a striated muscle fibre; they play a role in triggering the opening of voltage dependent Ca^{2+} channels in the smooth endoplasmic reticulum

tubulin: a protein found in microtubules

tyrosine kinase: an enzyme that adds phosphate groups derived from ATP to tyrosine residues in protein molecules

uniport: a membrane protein that transports a molecule from one side of the membrane to the other down a concentration gradient

uracil: a pyrimidine base found in RNA; it is equivalent to thymine in DNA and pairs with adenine

urea: a molecule $(CO(NH_2)_2)$ which is excreted in urine; it is used to remove waste nitrogen from the body

van der Waals forces: weak, short-range, inter- or intra-molecular attractions or repulsions between non-polar atoms

vein: a blood vessel which carries blood to the heart

ventricle: a large chamber of the heart responsible for pumping blood to the lungs (right ventricle) or the rest of the body (left ventricle)

virus: an infectious agent consisting of nucleic acid and protein that must infect a host cell in order to replicate

voltage sensitive channel: an ion channel that opens in response to a change in membrane potential

white matter: that part of the central nervous system containing myelinated axons

Z-line: a feature of striated muscle; a protein band into which actin filaments at either end of the sarcomere are inserted

zona pellucida: the coating surrounding the oocyte which the sperm must penetrate for fertilisation to occur

zygote: a diploid cell formed after fusion of an egg and sperm

Index